Molecular Beacons

Chaoyong James Yang • Weihong Tan
Editors

Molecular Beacons

Springer

Editors
Prof. Chaoyong James Yang
Department of Chemical Biology
College of Chemistry and Chemical Engineering
Xiamen University
Xiamen
P.R. China

Prof. Weihong Tan
Department of Biomedical Engineering and Department of Chemistry
Hunan University
Changsha
P.R. China

Department of Chemistry
University of Florida
Gainesville, FL, USA

ISBN 978-3-642-39108-8 ISBN 978-3-642-39109-5 (eBook)
DOI 10.1007/978-3-642-39109-5
Springer Heidelberg New York Dordrecht London

Library of Congress Control Number: 2013945903

Printed on acid-free paper

Springer is part of Springer Science+Business Media (www.springer.com)

Foreword

The detection of particular nucleic acid sequences is important for scientists in a number of disciplines. Cell biologists need to see where in the cell particular RNAs are present; pathologists need to detect cancerous cells and infectious organisms by virtue of the characteristic nucleic acid sequences that they possess; and clinical biologists wish to isolate stem cells, based on their unique gene expression profiles. In 1996, our laboratory introduced a simple method for nucleic acid detection by developing molecular beacon probes. Molecular beacons are synthetic DNAs that possess a quenched fluorophore that becomes brightly fluorescent when the probes hybridize to complementary DNA or RNA sequences.

Thanks to the efforts of many investigators, particularly Weihong Tan and Chaoyong James Yang, the authors of this remarkable book, that simple idea has found applications in fields far beyond what was originally imagined. Molecular beacons are now employed worldwide by a vibrant clinical diagnostics industry; and they are incorporated into diverse assays for the detection of many pathogens, including *Staphylococcus aureus, Mycobacterium tuberculosis*, the human immunodeficiency virus HIV-1, and human papilloma viruses.

Tan and Yang capture the excitement that pervades the hundreds of laboratories around the world that have developed and exploited molecular beacons technology. They begin by describing the design, synthesis, and purification of molecular beacons; and they then explore their thermodynamic and kinetic properties, and their use in real-time PCR measurements, in the detection of single-nucleotide polymorphisms, in applications where the probes are tethered to solid surfaces, and in the imaging of mRNAs in living cells. Furthermore, one of the most exciting developments that the authors describe is their laboratory's development of aptamer molecular beacons, which can be used to detect proteins, and they describe how these aptamers can be employed for the selective in vivo imaging of tumor cells.

The fruits of many years of molecular beacons research, in both academic and commercial laboratories, are scattered throughout the scientific literature. Tan and Yang provide an organized and comprehensive view of this literature. This exceptionally well-written and authoritative compendium will stimulate the imagination of researchers in diverse fields ranging from analytical chemistry to cell biology, and will provide a detailed foundation for the development of significant new applications in the clinical diagnostics community.

Public Health Research Institute
Rutgers University
Newark, NJ, USA

Sanjay Tyagi
Fred Russell Kramer

Contents

Contributors

Zehui Cao Dow AgroSciences LLC, Indianapolis, IN, USA

Tao Chen Department of Chemistry, University of Florida, Gainesville, FL, USA

Da Han Department of Chemistry, University of Florida, Gainesville, FL, USA

Jin Huang State Key Laboratory of Chemo/Biosensing and Chemometrics, College of Chemistry and Chemical Engineering, Hunan University, Changsha, China

Colin D. Medley Small Molecule Pharmaceutical Sciences, Genentech, USA

Lu Peng Department of Chemistry, University of Florida, Gainesville, FL, USA

Weihong Tan Department of Biomedical Engineering and Department of Chemistry, Hunan University, Changsha, P.R. China

Department of Chemistry, University of Florida, Gainesville, FL, USA

Chunming Wang College of Chemistry and Chemical Engineering, Xiamen University, Xiamen, People's Republic of China

Cuichen Wu Department of Chemistry, University of Florida, Gainesville, FL, USA

Chaoyong James Yang Department of Chemical Biology, College of Chemistry and Chemical Engineering, Xiamen University, Xiamen, P.R. China

Mingxu You Department of Chemistry, University of Florida, Gainesville, FL, USA

Zhi Zhu Department of Chemical Biology, College of Chemistry and Chemical Engineering, Xiamen University, Xiamen, P.R. China

Chapter 1
Design, Synthesis, Purification, and Characterization of Molecular Beacons

Zhi Zhu

Abstract Molecular beacon is hairpin-structured single-stranded oligonucleotide that can report the presence of specific nucleic acids through hybridization. It has been widely used for nucleic acid detection with high sensitivity and selectivity. The excellent performance of molecular beacons is primarily dependent on ingenious design, successful synthesis, as well as reliable purification. In this chapter, the design principle of molecular beacon is first presented, including the design of loop and stem as well as the choice of fluorophore and quencher pair. Then, the standard protocols of molecular beacon synthesis and purification are introduced. Finally, the characterization method of molecular beacon has to be done to ensure the production of high-quality MB probes and, subsequently, their implementation in accurate and reliable assays and applications.

1.1 Introduction

The hybridization of a nucleic acid strand to its complement target is one of the most specific molecular recognition events known. First reported by Tyagi and Kramer in 1996 [1], molecular beacons, commonly termed molecular beacon probes, are based on this principle. Molecular beacons (MBs) are oligonucleotide hybridization probes that can report the presence of specific nucleic acids in homogeneous solutions. This single-stranded DNA molecule consists of a stem-and-loop structure doubly labeled with a fluorophore and a quencher group on each end (F and Q, respectively, in Fig. 1.1). As shown in Fig. 1.1, MBs act like switches that are normally closed by the stem part, and in the "off" position, little fluorescence background is noted by the effect of quenching. However, upon binding with

Z. Zhu (✉)
Department of Chemical Biology, College of Chemistry and Chemical Engineering, Xiamen University, Xiamen 361005, P.R. China
e-mail: zhuzhi@xmu.edu.cn

C.J. Yang and W. Tan (eds.), *Molecular Beacons*, DOI 10.1007/978-3-642-39109-5_1,

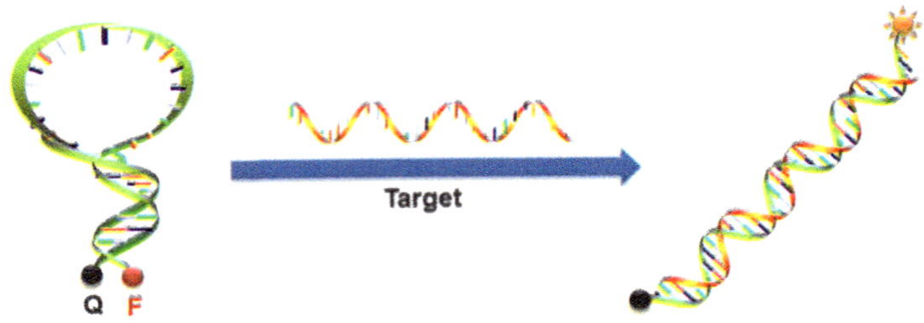

Fig. 1.1 Basic principle of molecular beacons

their targets, conformational changes open the hairpin, and fluorescence is turned "on" [2–6]. Molecular beacon probes are characterized by simple operation and high sensitivity and specificity. As such, they have become a class of molecules widely used in chemistry, biology, biotechnology, and biomedicine for biomolecular recognition [5, 7–9].

The excellent performance of MBs is primarily dependent on ingenious design, successful synthesis, as well as reliable purification. In this chapter, we will summarize the development of molecular beacons over the last 15 years, focusing on their design, synthesis, purification, and characterization. First, general design principles are introduced, including the choice of probe sequence, stem portion, and suitable fluorophore/quencher pair. Next, the synthesis and purification of MBs are described in detail. Both steps are crucial for the generation of high-quality probes. Finally, the basic methods of characterizing MBs are presented.

1.2 Design Principles of Molecule Beacons

A molecular beacon probe is a fluorescent-labeled oligonucleotide with 25–35 nucleotides long. A typical molecular beacon structure can be divided into four parts:

- *Loop*: This is the 18–30 single-strand region of the MB which is complementary to the target sequence.
- *Stem*: The MB stem is formed by the attachment, to both termini of the loop, of two short (5–7 nucleotide residues) oligonucleotides that are complementary to each other.
- *5′ Fluorophore*: At the 5′-end of the MB, a fluorescent dye is covalently attached.
- *3′ Quencher (nonfluorescent)*: The quencher dye is covalently attached to the 3′ end of the MB. When the MB is in closed-loop shape, the quencher resides in proximity to the fluorophore, which results in quenching the fluorescent emission of the latter.

The design principle of each separate part follows.

1.2.1 Probe Sequence Design

The process of molecular beacon design begins with the selection of the probe sequence, the determinant of probe specificity. To begin, an amplicon is a piece of DNA formed as the product of natural or artificial amplification events. For example, it can be formed by polymerase chain reactions (PCR) or ligase chain reactions (LCR) as well as by natural gene duplication. If the MB is designed to detect synthetic products during PCR, then any region within the amplicon that is outside the primer binding sites can be selected [8, 10–12]. The probe sequence of the MB should be long enough to bind to its target at the annealing temperature. In order to discriminate between amplicons that differ from one another by as little as a single-nucleotide substitution, the length of the probe sequence should also be such that it dissociates from its target at temperatures 7–10 °C higher than the annealing temperature of the PCR. The melting temperature of the probe-target hybrid can be predicted using the "percent-GC" (guanine and cytosine) rule or "nearest-neighbor" rules [available in most probe or primer design software packages, such as OligoAnalyzer 3.1 (http://www.idtdna.com/analyzer/applications/oligoanalyzer/) [13] or NUPACK (www.nupack.org, developed by Caltech)] [14]. The prediction should be made for the probe sequence alone before choosing the stem sequences.

If, on the other hand, single-nucleotide allele discrimination is not desired, longer and more stable probes can be chosen. However, while increase in the probe length results in improved affinity, it leads to reduced specificity. Nonetheless, the sensitivity and specificity of the probe for its target can be optimized for the desired application by adjusting the sequence length and GC content. As a general rule, probe sequences should be more than double the length of the stem portions in order to ensure the structure changes after hybridization, thus sufficiently separating the two signaling moieties. Generally, the length of the probe sequence falls in the range of 15–30 nucleotides and should not form any secondary structure [1, 11, 15]. If this happens, then the frame of the probe can be moved along the target to obtain a probe sequence that has minimal secondary structures. Small stems within the probe that are 2–3 nucleotides long do not adversely affect the performance of MBs.

1.2.2 Stem Sequence Design

After selecting the probe sequence, two complementary arm sequences are added on either side of the probe sequence. This is the stem sequence. The stem is designed in such a way that it should have a melting temperature of 7–10 °C more than the detection temperature [1, 16]. Usually, the stems are 5–6 base pairs and have a very high GC content (75–100 %) [8]. The length and the GC content of the stem sequence are designed in such a way that at the annealing temperature of the PCR, and in the absence of the target, the MBs remain closed and nonfluorescent. Since the stem is created by intramolecular hybridization, the melting temperature of the stem cannot be predicted by the percent-GC rule. Instead, a DNA folding program, such as the Zuker DNA folding program, mfold, and UNAfold available at http://mfold.rna.albany.edu/ [13], can be utilized to estimate the melting temperature of the stem. In general, 5-bp GC-rich stems have a melting temperature between 55 and 60 °C, 6-bp GC-rich stems have a melting temperature between 60 and 65 °C, and 7-bp GC-rich stems have a melting temperature between 65 and 70 °C [17]. MBs with shorter stem length have faster hybridization kinetics but suffer from lower signal-to-background ratio. In contrast, MBs with longer stem length can increase the difference of melting temperature between perfectly complementary duplex and mismatched duplex; therefore, MBs can have higher specificity but then slower hybridization kinetics [18–20]. Another concern is the final nucleotide positioned just before the fluorophore. Several investigators have observed that nucleotides can quench the fluorescence emission of fluorophores [4, 21]. While the quenching efficiency of the bases is smaller than those of commonly used quenchers, it can still significantly affect the performance of the MB. Guanine (G) is the best quencher among the bases, followed closely by adenine (A), while cytosine (C) and thymine (T) exhibit a much lower quenching efficiency. The quenching properties of guanosine are centered on its electron donating ability that allows the energy transfer between the base and the fluorophore. Fluorophores with green and yellow emission were quenched more efficiently by nucleotides than fluorophores with blue and red emission. As a result, care must be taken in the stem design to avoid the direct influence of neighboring nucleotide to the fluorophore.

After designing of both probe and stem, it is important to confirm that the MB does, indeed, form a hairpin structure in the absence of target. If the fluorophore is not close to the quencher, it will cause high background. If the stem is longer than expected, it will slow down the kinetics of hybridization with the target, an issue described at length in Chap. 2. If unexpected secondary structures result from the choice of the stem, a different stem sequence needs to be designed. The use of DNA folding software can prevent these problems.

1.2.3 Energy Transfer Mechanism for Molecular Beacons

Essentially, two forms of energy transfer may exist in MBs: dynamic and static fluorescence quenching [22].

Dynamic quenching includes Förster transfer (RET or FRET) and Dexter transfer (collisional quenching or electron-transfer quenching) [22]. It can occur between the donor and the acceptor by a nonradiative, long-range dipole-dipole coupling mechanism. Three basic requirements need to be satisfied: (1) that the distance between the donor and the acceptor be within 20–100 Å, as the two moieties brought any closer will result in reduced intensity; (2) that the spectral overlap be between the donor's emission and the acceptor's absorption spectra; and (3) that the relative orientation of the donor and acceptor be approximately parallel. The rate of FRET is inversely proportional to the sixth power of the separation distance of the donor and the acceptor or quencher.

Static or contact quenching requires much closer contact between the two moieties and the formation of ground-state complex, an intramolecular dimer with its own unique properties, such as being nonfluorescent and having a unique absorption spectrum [16, 21–23]. Dye aggregation often occurs through hydrophobic effects when the dye molecules stack together to minimize contact with water. Planar aromatic dyes that are matched for association through hydrophobic, electron steric forces can enhance static quenching. High temperatures and addition of surfactants tend to disrupt ground-state complex formation. Here, spectral overlap is not a significant determinant for quenching, and quenching efficiencies remain very high in the absence of spectral overlap. The collision between the fluorophore and the quencher distorts the energy level of the excited fluorophore, which, in turn, causes quenching. Most of the absorbed energy from fluorophore is dissipated as heat, rather than emitted as light. Both forms of energy transfer strongly depend on the distance between the two moieties on each terminus of the stem. Therefore, it is their spatial separation that determines the efficiency of energy transfer. When a target DNA hybridizes to an MB, fluorescence increases substantially with the larger separation between the two moieties. In most cases, static quenching, the original signaling mechanism of MBs, is the leading mechanism in MBs because of the short distances between the dyes [1].

1.2.4 Choice of Fluorophore and Quencher

The appropriate choice of fluorophore and quencher pair is critical for improved signal-to-background ratio. A systematic study of various static quenching efficiencies for different fluorophore and quencher pairs was summarized in Table 1.1 [4, 21]. The available organic dyes for MBs can cover the whole visible range from 400 to 700 nm. This flexibility in reporter dye is very useful in multiplex MB detection reactions [16, 24, 25]. For example, a multiplex nucleic acid assay simultaneously detecting four different retroviruses was developed with the use of different colored MBs [25]. Frequently used nonfluorescent quenchers include dabcyl (4-(dimethylaminoazo)benzene-4-carboxylic acid) and Black Hole Quenchers BHQ1 and BHQ2 [16, 21], all of which can serve as universal quenchers for a variety of fluorophores. The typical static quenching efficiency of these compounds lies between 80 and 98 %. Since the requirement of spectral overlap is not observed by the static quenching mechanism, different colored fluorophores can be quenched by the same nonfluorescent quenches. For example, BHQ-1 can quench the fluorophore from visible to near infrared, even though its maximum absorbance is 534 nm [4]. The appropriate fluorophore can be chosen on the basis of the instrumentation available for excitation and emission detection of fluorophore and the real-time application. Then, matching the fluorophore with an effective quencher can lead to improvements in detection by reducing the background from the MB in

Table 1.1 Static quenching efficiency of different fluorophore-quencher combinations [21]

Fluorophore	E_{max} (nm)	Dabcyl (A_{max} 475 nm)	BHQ-1 (A_{max} 534 nm)	QSY-7 (A_{max} 571 nm)	BHQ-2 (A_{max} 580 nm)
Alexa 350	441	95 %	97 %	97 %	96 %
Cy2	507	95 %	98 %	96 %	97 %
Alexa 488	517	94 %	95 %	95 %	93 %
FAM	517	91 %	93 %	93 %	92 %
Alexa 430	535	76 %	92 %	77 %	96 %
Alexa 532	551	93 %	95 %	96 %	93 %
Cy 3	564	94 %	97 %	95 %	93 %
Alexa 546	570	93 %	98 %	98 %	96 %
TMR	577	83 %	87 %	87 %	86 %
Cy 3.5	593	89 %	96 %	95 %	95 %
Alexa 568	599	91 %	98 %	99 %	98 %
Texas Red	603	96 %	98 %	98 %	97 %
Alexa 594	612	90 %	95 %	95 %	94 %
Alexa 633	645	96 %	98 %	97 %	97 %
Cy 5	663	84 %	96 %	79 %	96 %
Cy 5.5	687	82 %	96 %	74 %	95 %
Alexa 660	690	81 %	96 %	94 %	95 %
Alexa 680	702	81 %	94 %	90 %	93 %

E_{max} the emission maximum of the fluorophore, A_{max} absorption maximum of the quencher

the absence of targets. Researchers also observed that the affinity of fluorophore and quencher affects DNA hybridization [21]. For instance, the melting temperature of hybrids without any labels is 49 °C. However, with different fluorophore-quencher labeling, the melting temperatures range from 51 to 59 °C, which also correlates with the quenching efficiency. Fluorophore-quencher pairs that form strong bonding and stabilize the hybrid with higher melting temperatures have higher quenching efficiencies.

1.3 Synthesis of Molecular Beacons

In this section, the general procedure of DNA synthesis will be reviewed, in particular, the synthesis of molecular beacons.

1.3.1 Chemical Synthesis of Nucleic Acids

Currently, the chemical synthesis of relatively short nucleic acids (fewer than 200 bases) with a defined sequence is a remarkably simple, rapid, and inexpensive task on a commercial DNA synthesizer. For example, the Applied Biosystems 3400

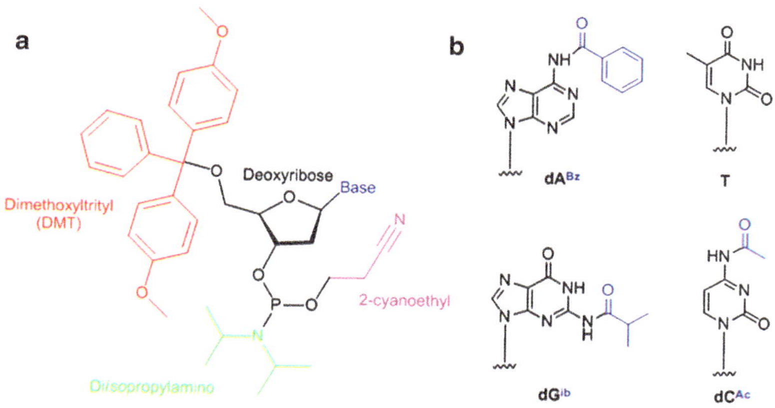

Fig. 1.2 Phosphoramidite chemistry. (**a**) Structure of phosphoramidite. (**b**) Four monomers of nucleic acid phosphoramidite

DNA Synthesizer is a versatile, four-column, benchtop instrument designed for 40-nmol, 200-nmol, and 1-μmol synthesis scale. Developments in nucleic acid research have spurred a dramatic expansion of the range of available modifications, either on the sequence ends or in the middle, which, in turn, greatly broadens nucleic acid applications. Generally, these modifiers are available at www.glenresearch.com, www.chemgenes.com, www.biosearchtech.com, and others. Currently, the process is implemented as solid-phase synthesis using the phosphoramidite method and phosphoramidite building blocks derived from protected 2′-deoxynucleosides (dA, dC, dG, and T), ribonucleosides (A, C, G, and U), or chemically modified nucleosides, e.g., Locked Nucleic Acid (LNA).

The building block, phosphoramidite (Fig. 1.2a), is composed of different functional groups: a nucleoside base plus a sugar ring, a phosphate, and protection groups. A di*i*sopropyl phosphoramidite group is attached to the 3′-hydroxyl of a nucleoside, ensuring efficient coupling. The 5′-hydroxyl of deoxyribose is capped with an acid-labile *dimethoxytrityl* (DMT) group, which can be selectively activated under acidic conditions during the synthesis. To prevent undesired side reactions, the phosphite is protected by a 2-cyanoethyl group, and all primary amines of nucleosides have to be rendered unreactive by attaching specific protecting groups (Fig. 1.2b), all of which are base sensitive to enable effective removal by strong bases after synthesis. Other modifiers, such as fluorophores, amines, biotin, and spacers, although different in structure and protecting group, share the same strategy to design into functional phosphoramidites.

Whereas enzymes synthesize DNA and RNA in a 5′ to 3′ direction, chemical synthesis of oligonucleotide is carried out from 3′ to 5′. The synthesis cycle begins with a column containing the solid support (controlled pore glass bead (CPG)) with holes and channels in it, in which the first nucleotide on the 3′-end is covalently attached via its 3′-terminal hydroxyl group with a long space arm. The synthesis

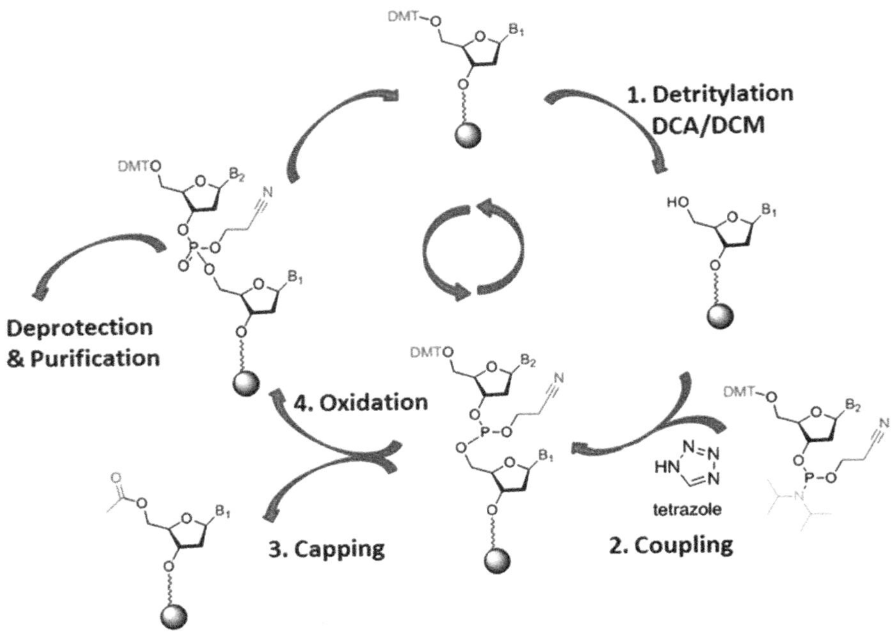

Fig. 1.3 Automated oligonucleotide synthesis through phosphoramidite chemistry

column has filters on both sides to hold the beads. The reagents pass through the filter to react with the beads, and excess reagents are easily removed. During synthesis, the phosphoramidite building blocks are sequentially coupled to the growing oligonucleotide chain on the beads.

One synthesis cycle involves four chemical reactions: detritylation, coupling, capping, and oxidation, as shown in Fig. 1.3.

The first step, detritylation, is a process whereby the column is flushed with a dilute acidic solution, either 3 % dichloroacetic acid (DCA) or trichloroacetic acid (TCA) in dichloromethane (DCM), to remove the trityl group protecting the 5′-hydroxyl at the end of the growing oligonucleotide chain attached to the CPG. Then the 5′-hydroxyl group becomes the only group on the column to react with the subsequent nucleotide. The intensity of the orange color from DMT liberated at this step is measured by a UV-vis detector inside the instrument to determine the coupling efficiency of the previous step.

The second step, coupling, is achieved by adding a phosphoramidite derivative of the next nucleotide with tetrazole, a weak acid, together through the column. The tetrazole protonates the diisopropylamine leading to the formation of the tetrazole-phosphoramidite intermediate, which is susceptible to nucleophilic attack by the active 5′-hydroxyl group on the CPG to form an unstable phosphate triester linkage. The reaction column is then washed to remove any extra tetrazole, unbound bases, and by-products.

Since the coupling yield is not always 100 %, some of the solid support-bound 5′-hydroxyl site remains unreacted and thus could react in later cycles, resulting in an oligonucleotide with a deletion. To prevent this from happening, the third step, capping, is needed to terminate or cap the unreacted 5′-hydroxyl groups by acetylation to become "failure products." This is achieved by co-delivering acetic anhydride and 1-methylimidazole through the column. The DMT group of the successful coupling step protects the 5′-hydroxyl group from being capped. This step can minimize the lengths of failure products and facilitate post-synthesis HPLC purification.

The last step, oxidation, converts the internucleotide linkage from the less stable phosphite to the much more stable phosphate linkage. It is achieved by treating the support-bound materials with iodine and water in presence of a weak base, such as pyridine. Iodine is used as the oxidizing agent and water as the oxygen donor.

1.3.2 Special Characteristics of Molecular Beacon Synthesis

Conventionally, MBs were synthesized by the manual coupling of fluorophore and quencher moieties to the oligonucleotides with amino and sulfhydryl groups at each end. However, the availability of various fluorophores and quenchers as phosphoramidite derivatives or linked to CPG solid supports has made it possible to synthesize most MBs using the protocol described above on a DNA synthesizer [2, 6]. For example, a CPG solid support, derivated with Dabcyl, BHQ-1, or BHQ-2, is used to start synthesis at the 3′-end. Then, the remaining nucleotides are added sequentially, as designed by standard phosphoramidite chemistry. At the 5′-end, a fluorophore phosphoramidite derivative is conjugated on the MBs. In most cases, no changes in DNA synthesis protocol are required in order to incorporate the 5′ fluorophores; however, it is highly recommended that the users refer to the instruction manuals provided by the manufacturers for special coupling and post-synthesis protocol for each phosphoramidite derivative.

Sometimes, a fluorophore or quencher is not available for direct incorporation on a DNA synthesizer. Chemically reactive fluorophore derivatives can still be introduced by a post-DNA synthesis step [12]. In this case, an amino or sulfhydryl phosphoramidite is incorporated at the 5′-end of the MB. Following a second post-DNA synthesis step, succinimidyl ester derivatives of fluorophores are coupled with an amino group, and either fluorophore iodoacetamide or maleimide derivatives are coupled to a sulfhydryl moiety. If the protecting group on amino and sulfhydryl can be easily removed without influencing other protecting groups on the oligonucleotide, this solid-liquid reaction, which has a higher yield, can directly take place on the CPG solid support. Otherwise, the oligonucleotide is first removed from the CPG and purified by high-pressure liquid chromatography (HPLC). Then, the fluorophore coupling reaction happens in liquid phase, and a second HPLC is needed.

After synthesis and modification are complete, the fully protected products are hydrolyzed, removed from the CPG solid support, and deprotected. The method varies according to the functional groups on the oligonucleotides. Normally, this is done by incubating the CPG in concentrated ammonia at high temperature for an extended amount of time. However, MBs require special attention. Particularly, the deprotection procedures will vary among manufacturers; thus, the manufacturer's instructions should be followed closely. Otherwise, an inappropriate deprotection method might irreversibly damage the fluorophore. With this step, all the protecting groups on the nucleosides and the 2-cyanoethyl group on the phosphate are removed. If, however, DMT is applied as the hydrophobic group to interact with HPLC stationary phase, only the DMT group on 5′-hydroxyl remains. Next, during incubation in ethanol at −20 °C for 30 min, the oligonucleotide and some salts precipitate from the solution, and, finally, the precipitate is collected by centrifugation at high speed and is ready for further purification.

1.4 Purification of Molecular Beacons

As described in Sect. 1.3, oligonucleotide chemical synthesis typically yields the target oligonucleotide contaminated with truncated by-products. Low-quality oligonucleotides may have a detrimental impact on performance, especially for MBs. Background fluorescence from failure products contributes to false-positive or false-negative results. Consequently, purification is the next important step for further assays and applications.

Traditionally, two techniques are widely used for DNA separation: polyacrylamide gel electrophoresis (PAGE) and chromatography [26–28]. Although the separation ability of PAGE is excellent for oligonucleotides within 100 mers with greater than 98 % purity in most cases, the method is lengthy and laborious. Also, it is difficult to scale up the oligonucleotide mass load without a loss of resolution. Moreover, this technique is difficult to automate.

The two most commonly used chromatographic modes are anion exchange and reversed phase. Anion exchange chromatography separates oligonucleotides based on the number of charges (phosphate linkages) and is very efficient for lengths from 2 to 30 mers with purity of about 95–98 % [29]. It should be noted that selectivity of the separation decreases as the length of the oligonucleotide increases, and sometimes this causes overall yield to be sacrificed in order to obtain a high purity product. Reversed-phase HPLC has become a popular technique for oligonucleotide purification with the introduction of the "trityl-on" (DMT-on) method, which is the most suitable technique for modified oligonucleotides, especially MB purification.

Actually, the most commonly used HPLC technique for oligonucleotide purification is reversed-phase ion-pairing HPLC (RP-IP HPLC) [26–28, 30]. In RP-HPLC, the alkyl chains interact with analytes and reverse the elution order. Here, the method utilizes hydrophobic properties of the dimethoxytrityl (DMT) protection moiety used in DNA synthesis. The DMT group is not cleaved from the 5′-end

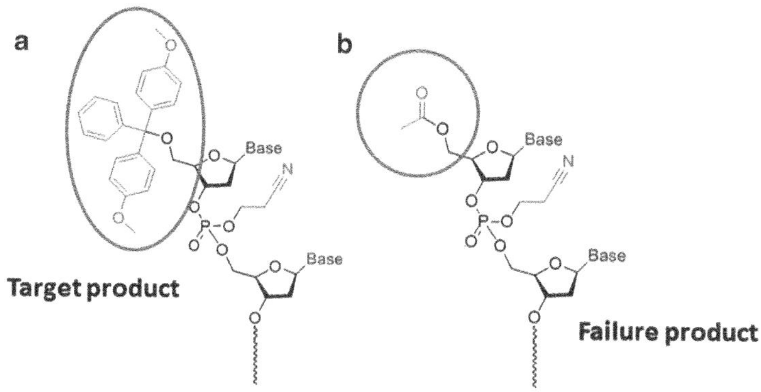

Fig. 1.4 Partial structure of a synthetic oligonucleotide. (**a**) "DMT-on" target product. (**b**) "DMT-off" failure product

of the full-length product after the very last synthesis step to assist in RP-HPLC, which is named "DMT-on" purification. The resulting synthetic mixture contains the full-length "DMT-on" target products and shorter "DMT-off" failure products (Fig. 1.4). The target product bearing the DMT group is strongly retained on RP-HPLC sorbents and elutes significantly later than the failure product. The separation selectivity is greatly enhanced by the presence of DMT group, allowing for convenient purification of oligonucleotides up to about 100 mer.

Meanwhile, the stationary phase, alkyl chains on column, is nonionic and hydrophobic. However, DNA molecule is ionic due to the phosphate backbone and, without any treatment, interacts poorly with the stationary phase. In order to convert DNA from polar to nonpolar, ion-pairing chromatography (IP-HPLC) is introduced. An additive organic cation, as part of the eluent, allows the nucleic acid to become nonpolar. Each anionic phosphate group of DNA is paired with an organic cation, as shown in Fig. 1.5a. The organic cation is a positively charged ammonium cation that also contains alkyl groups. This interaction results in the formation of an ion pair that is more "organic" than a normal oligonucleotide and can therefore be separated on a RP-HPLC column with an organic-aqueous mobile phase such as acetonitrile (ACN) water.

In DNA chromatography, a fairly small alkyl group is used as the organic cation. The most common one is triethylammonium (TEA) cation. To prepare this, acetic acid is added to triethylamine to make a final solution with a pH of 7. The solid DNA is redissolved in 0.1-M triethylammonium acetate (TEAA, pH 7) for RP-IP HPLC with a C_{18} column as stationary phase and ACN and 0.1-M TEAA water as the mobile phase. The purification starts with pumping 0.1-M TEAA through the column until equilibrium is attained. Then the DNA in TEAA is injected and absorbs to the stationary phase. A gradient of ACN is started. As a typical HPLC gradient, a linear elution gradient of 10–60 % buffer B (ACN) in buffer A (0.1-M TEAA) forms over 25 min at a flow rate of 1 mL/min (blue line) (see Fig. 1.5b) [17]. Normally, a UV-vis detector is equipped with HPLC to monitor the DNA peak at 260 nm and

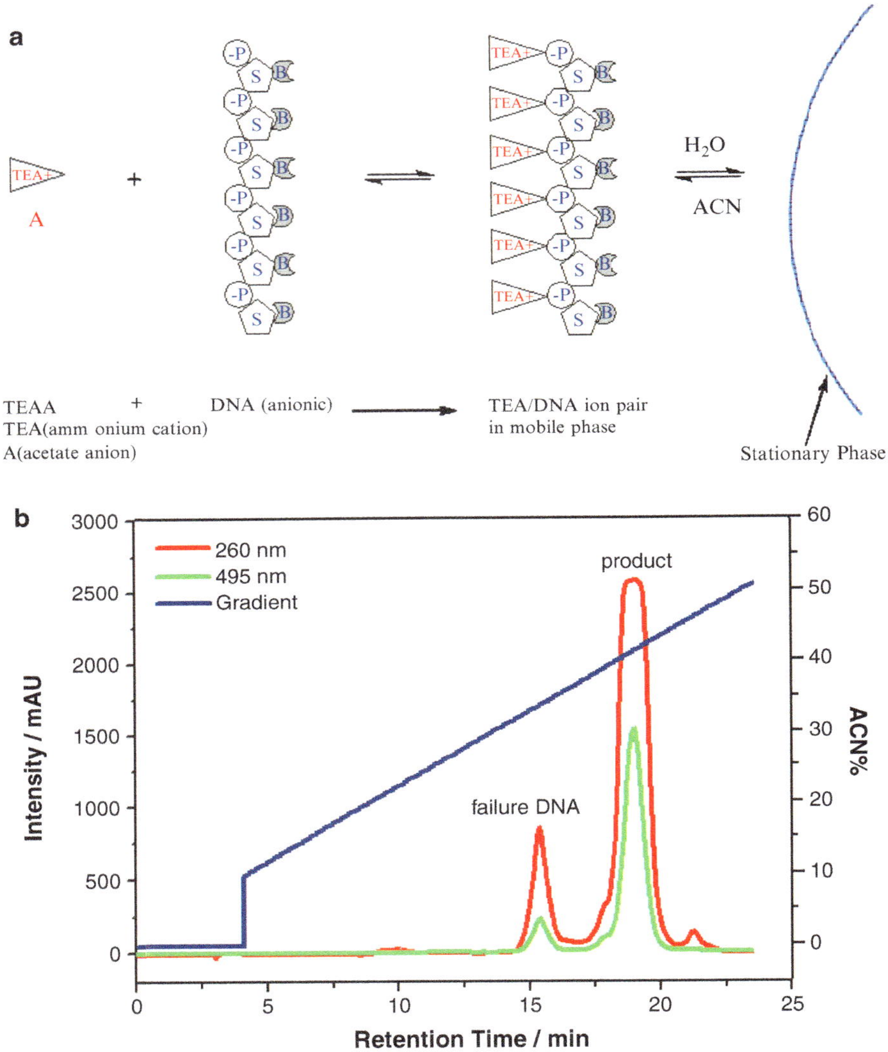

Fig. 1.5 HPLC purification of MBs. (**a**) The reversed-phase ion-pairing process. The negatively charged nucleic acid is combined with the positively charged TEA ion to form the neutral ion pair which can interact with the column's surface. *TEA* triethylammonium, *A* acetate, *P* phosphate, *S* sugar, *B* base. (**b**) A typical HPLC gradient and results for MB purification. *Blue line* shows the gradient of acetonitrile from 10 to 50 % over 20 min. The *red* and *green lines* monitor the separation process simultaneously through a DNA peak at 260 nm and a fluorophore/quencher peak around 495 nm, respectively

the fluorophore/quencher absorbance around 495 nm. First, the "DMT-off" failure products are desorbed and travel down the column to the detector as the first peak of red and green lines, as shown in Fig. 1.5b. Then, as the organic solvent content of the eluent is increased, "DMT-on" target products are desorbed and also travel down the column to give the second peak. Finally, after all the fragments of interest are eluted, the column is cleaned of residual material with a final gradient concentration of high ACN. After this, the eluent is returned to original eluent concentrations, and the column is conditioned for the next run [30].

In certain cases, instead of a DMT group, a quencher molecule with similar hydrophobicity is coupled on the 5′-end. By slightly adjusting the gradient according to each practical situation, the principle and process of HPLC separation is similar to that with DMT group. Thus, to better separate product with failure DNA, the 5′ modification with a hydrophobic quencher is recommended. Sometimes, dual HPLC is performed as well to obtain high product purity.

After the purified oligonucleotide is dried, the DMT group is removed by incubation with 80 % acetic acid. The reaction is terminated with ethanol followed by vacuum drying. The pure oligonucleotide can be quantified by UV absorbance measurement at 260 nm.

1.5 Characterization of Molecular Beacons

The newly synthesized MB can be characterized by mass spectra and its structure confirmed by molecular weight. Importantly, MBs can also be characterized according to signal-to-background ratio and their thermal denaturation profile [5, 9, 12].

1.5.1 Evaluation of Signal-to-Background Ratio

In order to ensure that an MB will function as expected, the extent to which its fluorescence increases upon binding to its target should first be measured. The signal-to-background ratio depends on both the design and the purity of MB preparation. A poor signal-to-background ratio is the primary cause for the impurity of MBs, either by the presence of uncoupled fluorophores in the preparation or the presence of oligonucleotides containing a fluorophore but lacking a quencher.

The signal-to-background ratio of an MB can be determined by using a spectrofluorometer. All measurements are taken at the same temperature and buffer condition as the planned experiment, and the appropriate excitation and emission source settings are selected for the fluorophore that is used as a label for the MB. The fluorescence of buffer solution is first determined as F_{buffer}. A certain concentration of MB is added, normally around 50–200 nM, since self-quenching would be significant at too high concentration. The fluorescence is measured as F_{closed}. Then, fivefold molar excess of oligonucleotide target, whose sequence is

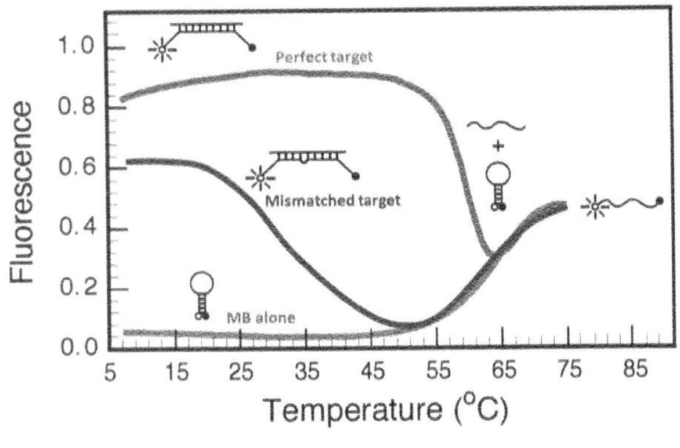

Fig. 1.6 The illustrative thermal denaturation profiles of MBs in the absence of target (*green line*), in the presence of either perfect target (*red line*) or mismatched target (*blue line*) (Color figure online)

perfectly complementary to the probe sequence of MB, is added, and the rise in fluorescence is monitored until it reaches a stable level as F_{open}. The signal-to-background ratio is calculated as $(F_{open} - F_{buffer})/(F_{closed} - F_{buffer})$. Preferably, the MBs should have signal-to-background ratios above 20.

1.5.2 Determination of Thermal Denaturation Profiles

Temperature also plays an important role [8, 9, 17]. In order to determine the window of discrimination, it is important to understand how the fluorescence of MBs changes with temperature in the absence and in the presence of both perfect targets and mismatched targets [20]. In the presence of each kind of target, a thermal cycler is used to monitor the fluorescence of solutions of MBs as a function of temperature. The temperature of these tubes is decreased from 80 to 10 °C in 1 °C increments, with each hold lasting 1 min, while monitoring the fluorescence during each hold.

In the absence of targets, as shown by the green fluorescence versus temperature trace in Fig. 1.6, MBs, at low temperature, can maintain a stable hairpin structure in the closed (*off*) state, where the fluorophore and the quencher are held in close proximity to each other and no fluorescence emits. However, at high temperature, MBs will extend to become a random-coil configuration, thus separating fluorophore and quencher, consequently emitting fluorescence. MB melting temperature depends on the chain length of the stem, GC content, as well as the ionic concentration of the buffer.

In the presence of targets, MBs have three possible phases: hybridized with a target, free in the stem-loop conformation, and free as a random coil [20]. As shown by the red trace in Fig. 1.6, at low temperature, the MB spontaneously binds to its perfect target, forming a stable hybrid and turning on its fluorescence. With the gradual increase of temperature, the MB dissociates from its target and returns to its hairpin structure, significantly diminishing the fluorescence. The temperature at which the MB-target hybrid melts apart depends upon the GC content and the length of the probe sequence. By increasing the temperature, the MB's hairpin structure dissolves and is replaced by a random-coil structure. The fluorophore and quencher then separate, and fluorescence recovers. Meanwhile, the fluorescence curve of an MB with its mismatched target is shown as the blue trace in Fig. 1.6. The dissociation temperature of MB and mismatched target is much lower than that of MB and perfect target.

The thermal denaturation profile establishes the working temperature range of an MB such that perfectly complementary target-probe hybrids can form, but mismatched target-probe hybrids cannot. The selectivity potential of a given MB is also determined by the difference between these phase-transition temperatures.

1.6 Conclusion

While the content of this chapter has presented only the fundamentals of MB design, synthesis, purification, and characterization, each plays a critical role in ensuring the production of high-quality MB probes and, subsequently, their implementation in accurate and reliable assays and applications. Looking ahead, some artificial MBs have also been designed by replacing natural DNA with special nucleotides, such as LNA, L-DNA, 2-OMe RNA, or PNA, which can reduce nonspecific interactions and extend the lifetime in vivo [31–34]. Some MBs have also been designed with super-quenchers, and some are bioconjugates, being linked to nanomaterials and polymers [35, 36]. These types of MBs still follow the basic principles discussed and will be explained in detail in later chapters.

References

1. Tyagi S, Kramer FR (1996) Molecular beacons: probes that fluoresce upon hybridization. Nat Biotechnol 14(3):303–308
2. Fang XH, Li JWJ, Perlette J, Tan WH, Wang KM (2000) Molecular beacons – novel fluorescent probes. Anal Chem 72(23):747a–753a
3. Tan WH, Wang KM, Drake TJ (2004) Molecular beacons. Curr Opin Chem Biol 8(5):547–553
4. Drake TJ, Tan W (2004) Molecular beacon DNA probes and their bioanalytical applications. Appl Spectrosc 58(9):269A–280A
5. Wang K, Tang Z, Yang CJ, Kim Y, Fang X, Li W, Wu Y, Medley CD, Cao Z, Li J, Colon P, Lin H, Tan W (2009) Molecular engineering of DNA: molecular beacons. Angew Chem Int Ed Engl 48(5):856–870
6. Tan WH, Fang XH, Li J, Liu XJ (2000) Molecular beacons: a novel DNA probe for nucleic acid and protein studies. Chem-Eur J 6(7):1107–1111
7. Tyagi S (2009) Imaging intracellular RNA distribution and dynamics in living cells. Nat Methods 6(5):331–338
8. Singh K, Goel G, Kumar A, Puniya AK, Chen W (2005) Molecular beacon: a multitask probe. J Appl Microbiol 99(3):435–442

9. Ye DY, Li YS, Zhou XY (2008) Molecular beacons: an optimal multifunctional biological probe. Biochem Biophys Res Commun 373(4):457–461
10. Broude NE (2002) Stem-loop oligonucleotides: a robust tool for molecular biology and biotechnology. Trends Biotechnol 20(6):249–256
11. Ortiz E, Lizardi PM, Estrada G (1998) PNA molecular beacons for rapid detection of PCR amplicons. Mol Cell Probe 12(4):219–226
12. Vet JAM, Marras SAE (2004) Chapter 17: Design and optimization of molecular beacon real-time polymerase chain reaction assays. In: Methods in molecular biology, vol 288: Oligonucleotide synthesis: methods and applications. Humana Press Inc, Totowa, pp 273–289
13. Zuker M (2003) Mfold web server for nucleic acid folding and hybridization prediction. Nucleic Acids Res 31(13):3406–3415
14. Pierce NA, Zadeh JN, Steenberg CD, Bois JS, Wolfe BR, Pierce MB, Khan AR, Dirks RM (2011) NUPACK: analysis and design of nucleic acid systems. J Comput Chem 32(1):170–173
15. Kostrikis LG, Tyagi S, Mhlanga MM, Ho DD, Kramer FR (1998) Molecular beacons – spectral genotyping of human alleles. Science 279(5354):1228–1229
16. Tyagi S, Bratu DP, Kramer FR (1998) Multicolor molecular beacons for allele discrimination. Nat Biotechnol 16(1):49–53
17. http://www.molecular-beacons.org/
18. Libchaber A, Bonnet G, Krichevsky O (1998) Kinetics of conformational fluctuations in DNA hairpin-loops. Proc Natl Acad Sci U S A 95(15):8602–8606
19. Bao G, Tsourkas A, Behlke MA, Rose SD (2003) Hybridization kinetics and thermodynamics of molecular beacons. Nucleic Acids Res 31(4):1319–1330
20. Kramer FR, Bonnet G, Tyagi S, Libchaber A (1999) Thermodynamic basis of the enhanced specificity of structured DNA probes. Proc Natl Acad Sci U S A 96(11):6171–6176
21. Marras SA, Kramer FR, Tyagi S (2002) Efficiencies of fluorescence resonance energy transfer and contact-mediated quenching in oligonucleotide probes. Nucleic Acids Res 30(21):e122
22. Lakowicz JR (2006) Principles of fluorescence spectroscopy, 3rd edn. Springer Science + Business Media LLC, New York
23. Bernacchi S, Mely Y (2001) Exciton interaction in molecular beacons: a sensitive sensor for short range modifications of the nucleic acid structure. Nucleic Acids Res 29(13):E62–62
24. Tyagi S, Marras SAE, Kramer FR (2000) Wavelength-shifting molecular beacons. Nat Biotechnol 18(11):1191–1196
25. Kramer FR, Vet JAM, Majithia AR, Marras SAE, Tyagi S, Dube S, Poiesz BJ (1999) Multiplex detection of four pathogenic retroviruses using molecular beacons. Proc Natl Acad Sci U S A 96(11):6394–6399
26. Gilar M (2001) Analysis and purification of synthetic oligonucleotides by reversed-phase high-performance liquid chromatography with photodiode array and mass spectrometry detection. Anal Biochem 298(2):196–206
27. LaCourse WR, Gelhaus SL (2005) Separation of modified 2′-deoxyoligonucleotides using ion-pairing reversed-phase HPLC. J Chromatogr B 820(2):157–163
28. Gilar M, Fountain KJ, Budman Y, Neue UD, Yardley KR, Rainville PD, Russell RJ, Gebler JC (2002) Ion-pair reversed-phase high-performance liquid chromatography analysis of oligonucleotides: retention prediction. J Chromatogr A 958(1–2):167–182
29. Oefner PJ, Huber CG (2002) A decade of high-resolution liquid chromatography of nucleic acids on styrene-divinylbenzene copolymers. J Chromatogr B 782(1–2):27–55
30. Gjerde DT, Hanna CP, Hornby D (2003) Chromatographic principles for DNA separation, in DNA chromatography. Wiley-VCH Verlag GmbH & Co KGaA, Weinheim
31. Wang L, Yang CYJ, Medley CD, Benner SA, Tan WH (2005) Locked nucleic acid molecular beacons. J Am Chem Soc 127(45):15664–15665
32. Kim Y, Yang CJ, Tan W (2007) Superior structure stability and selectivity of hairpin nucleic acid probes with an L-DNA stem. Nucleic Acids Res 35(21):7279–7287
33. Yang CJ, Wang L, Wu Y, Kim Y, Medley CD, Lin H, Tan W (2007) Synthesis and investigation of deoxyribonucleic acid/locked nucleic acid chimeric molecular beacons. Nucleic Acids Res 35(12):4030–4041

34. Wu Y, Yang CJ, Moroz LL, Tan W (2008) Nucleic acid beacons for long-term real-time intracellular monitoring. Anal Chem 80(8):3025–3028
35. Yang CYJ, Lin H, Tan WH (2005) Molecular assembly of superquenchers in signaling molecular interactions. J Am Chem Soc 127(37):12772–12773
36. Yang CYJ, Pinto M, Schanze K, Tan WH (2005) Direct synthesis of an oligonucleotide-poly(phenylene ethynylene) conjugate with a precise one-to-one molecular ratio. Angew Chem Int Ed 44(17):2572–2576

Chapter 2
Thermodynamic and Kinetic Properties of Molecular Beacons

Lu Peng and Weihong Tan

Abstract Molecular beacons are widely used for detection of nucleic acids both in vitro and in vivo. Compared with linear probes, molecular beacons have shown enhanced sensitivity and specificity primarily due to their stem–loop hairpin structures. The hairpin structures bring new considerations on thermodynamics and kinetics for designing of nucleic acid probes. This chapter has been designed to provide a better understanding of structure–performance relationship of molecular beacons based on analysis of their thermodynamic and kinetic properties. The conformational fluctuations of molecular beacons are discussed concerning the stability and kinetics of the hairpin-coil transformation. In the presence of target nucleic acids, molecular beacons hybridize with targets to form duplex complexes. We analyzed the theoretical models and the relevant parameters used to describe the hybridization reactions. Furthermore, studies on strategies for optimization of molecular beacon performance are summarized. The systematic analysis of studies about thermodynamic and kinetic properties of molecular beacons allows for sophisticated design of better molecular beacons for specific purposes.

2.1 Introduction

Molecular beacons are designed with a target-specific hybridization domain positioned between two short self-complementary sequences [1]. In the absence of target, the intramolecular hybridization reaction between the self-complementary

L. Peng (✉)
Department of Chemistry, University of Florida, Gainesville, FL 32611, USA
e-mail: lupeng@ufl.edu

W. Tan
Department of Biomedical Engineering and Department of Chemistry, Hunan University, Changsha, P.R. China

Department of Chemistry, University of Florida, Gainesville, FL, USA
e-mail: tan@chem.ufl.edu

C.J. Yang and W. Tan (eds.), *Molecular Beacons*, DOI 10.1007/978-3-642-39109-5_2,

domains generates a stem–loop hairpin structure and brings the fluorophore into close proximity with the quencher, resulting in fluorescence quenching. For any applications using molecular beacons, the best probe performance depends on design optimization. For example, while high specificity is required to identify single nucleotide polymorphisms (SNP), real-time study of transient RNA expression in vitro requires fast hybridization kinetics [2]. Thus, knowledge of the thermodynamics and kinetics of molecular beacons is a design prerequisite. During the last several years, many studies have addressed the conformational fluctuation of molecular beacons and their interactions with targets from the perspective of thermodynamics and kinetics, and in this chapter, these properties will be discussed.

2.2 Conformational Fluctuations of Molecular Beacons

Molecular beacons represent a class of hairpin nucleic acid probes with stem–loop structures. Although the signaling groups (fluorophores and quenchers) report hybridization events, the performance and function of molecular beacons as hybridization probes are essentially determined by the thermodynamic and kinetic properties of hairpin structures. Hairpins are the dominant secondary structure element in RNA. Certain sequences possess exceptionally high thermodynamic stability with the ability to determine such biological functions as enzymatic and anti-RNase activity.

The hairpin structures of molecular beacons are not static. They fluctuate between different conformations, particularly the open and closed states, as shown in Fig. 2.1 [3]. As a consequence of base pairing in the hairpin's stem region, the closed state has low enthalpy. On the other hand, the open state has high enthalpy by the large number of configurations that can be achieved by a single-stranded DNA (ssDNA). Thus, while the transition from the closed state to the open state requires an energy force great enough to break all the base pairs, collision of the two hairpin arms is necessary to close the hairpin, followed by the nucleation and the propagation of base pairing in the stem.

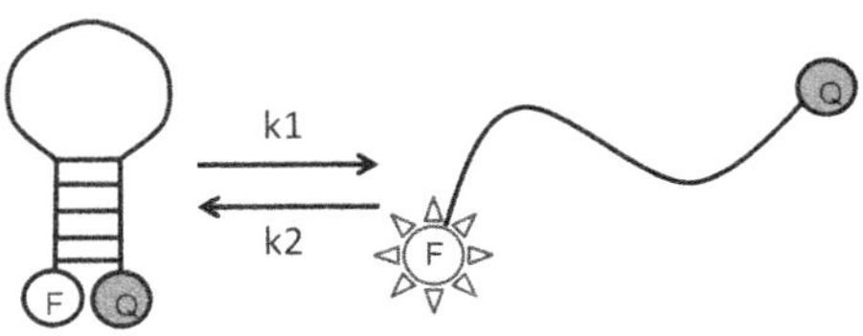

Fig. 2.1 Sketch of the conformational fluctuation of a DNA molecular beacon. The molecular beacon fluctuates between open and closed states with the rate constants k_1 and k_2. The fluorophore (F) and the quencher (Q) are covalently attached to the two arms. Fluorescence is quenched in the closed state; however, upon target recognition, the hairpin structure is opened, and the MB emits fluorescence

2.2.1 Thermodynamics of Hairpins

Enthalpy is a thermodynamic potential, measuring the total energy of a thermodynamic system. The total enthalpy, H, of a system cannot be measured directly. Thus, change in enthalpy, ΔH, is a more useful quantity than its absolute value. ΔH of a system is equal to the sum of nonmechanical work done on it and the heat supplied to it. Entropy is the thermodynamic property toward equilibrium/average/homogenization/dissipation. That is, the hotter, more dynamic areas of a system lose heat/energy, while cooler areas (e.g., space) get warmer/gain energy. Entropy of a system can be measured to determine the energy not available for work in a thermodynamic process.

Thermodynamic parameters of hairpins, including enthalpies and entropies, are revealed by melting studies [4–6]. In a typical experiment, the unfolding heat of a hairpin is measured with a differential scanning calorimeter (DSC). A DNA solution is scanned against a buffer solution within a certain temperature range, normalized by the heating rate and a buffer versus buffer scan subtracted and normalized by the effective number of moles used. Integration of the resulting curve, $\int \Delta C_{\mathrm{p}} \mathrm{d}T$, yields model-independent enthalpy, ΔH_{cal}. Change in entropy, ΔS_{cal}, is obtained by a similar integration, $\int (\Delta C_{\mathrm{p}}/T)\mathrm{d}T$, and checked against a standard relationship for intramolecular transitions: $\Delta S_{\mathrm{cal}} = \Delta H_{\mathrm{cal}}/T_{\mathrm{M}}$. In the measurement of these terms, it is assumed that the duplex and random-coil states have similar heat capacities. The free energy at any temperature T is obtained by the Gibbs relationship: $\Delta G^{\circ}_{\mathrm{cal}}(T) = \Delta H_{\mathrm{cal}} - T\Delta S_{\mathrm{cal}} = \Delta H_{\mathrm{cal}}\ (1 - T/T_{\mathrm{M}})$ [7].

The free energy of a hairpin can be broken into two parts: the free energy of forming a loop closed by a single base pair and the free energy for the base-paired stem of the hairpin. The free energy of the stem can be analyzed with the nearest-neighbor model for both stacking and base pairing [5, 8]. Therefore, the free energy for loop formation is determined by subtracting the free energy for stem from those measured free energy for the hairpin from optical or DSC melting studies.

2.2.2 Factors Affecting the Stability of Hairpins

The stability of a hairpin is characterized by its melting temperature. At the melting point, the change in Gibbs free energy (ΔG) of the material is zero, but the enthalpy (H) and the entropy (S) of the material are increasing (ΔH, $\Delta S > 0$). The melting phenomenon happens when the Gibbs free energy of the liquid becomes lower than the solid for that material. Accordingly, the relationship between melting temperature and Gibbs free energy is shown in the following equation: $\Delta G^{\circ}_{\mathrm{cal}}(T) = \Delta H_{\mathrm{cal}} - T\Delta S_{\mathrm{cal}} = \Delta H_{\mathrm{cal}}\ (1 - T/T_{\mathrm{M}})$. The factors that contribute to the difference in free energy are also responsible for the stability of hairpins.

The effect of stem structures on the stability of hairpins can be explained by the nearest-neighbor model. Briefly, the k-nearest-neighbor (k-NN) algorithm is among the simplest of all machine learning algorithms: an object is classified by a majority vote of its neighbors, with the object being assigned to the class most common among its k-nearest neighbors (k is a positive integer, typically small). If $k = 1$, then the object is simply assigned to the class of its nearest neighbor. In our case, the interaction between bases on the two arms of hairpins depends somewhat on the neighboring bases. Instead of treating a DNA duplex as a string of interactions between base pairs, the nearest-neighbor model treats a DNA helix as a string of interactions between "neighboring" base pairs [9]. As such, the enthalpy (ΔH°), entropy (ΔS°), and free energy (ΔG°) of duplex annealing in the stem region are predicted by the nearest-neighbor method and thermodynamic parameters [9, 10]. Stem length and the identity of the nearest-neighbor bases determine the stability of the stem. The arrows in the example below indicate the nearest-neighbor interactions in the stem of a hairpin.

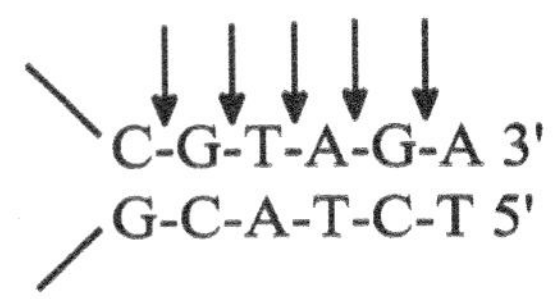

The free energy of forming the stem at 37 °C, ΔG°_{37} (stem), is represented as

$$\begin{aligned}\Delta G^{\circ}_{37}(\text{stem}) &= \Delta G^{\circ}_{37}(\text{CG initiation}) + \Delta G^{\circ}_{37}\left(\frac{\text{CG}}{\text{GC}}\right) + \Delta G^{\circ}_{37}\left(\frac{\text{GT}}{\text{CA}}\right)\\ &+ \Delta G^{\circ}_{37}\left(\frac{\text{TA}}{\text{AT}}\right) + \Delta G^{\circ}_{37}\left(\frac{\text{AG}}{\text{TC}}\right) + \Delta G^{\circ}_{37}\left(\frac{\text{GA}}{\text{CT}}\right)\\ &+ \Delta G^{\circ}_{37}(\text{AT initiation})\end{aligned} \tag{2.1}$$

The first term, ΔG°_{37} (CG initiation), represents the free energy of the first base pair, CG, in the absence of a nearest neighbor. The second term, ΔG°_{37} (CG/GC), includes both the free energy of base pairing, GC, and the stacking interaction of this base pair with the previous base pair, CG. The remaining terms are defined in a similar manner. As a general rule, the free energy of stem formation of a hairpin is calculated by the equation below:

$$\Delta G^{\circ}_{37}(\text{stem}) = \Delta G^{\circ}_{37}(\text{initiation}) + \sum_{i=1}^{10} n_i \Delta G^{\circ}_{37}(i) \tag{2.2}$$

Based on Gibbs relationship, ΔG°(stem) is also given by

$$\Delta G^{\circ}_{37}(\text{stem}) = \Delta H^{\circ}(\text{stem}) - T\Delta S(\text{stem}) \tag{2.3}$$

Table 2.1 Nearest-neighbor parameters for DNA/DNA duplexes in 1 M NaCl

Nearest-neighbor sequence (5′-3′/3′-5′)	ΔH° kJ/mol	ΔS° J/(mol · K)	ΔG° kJ/mol
AA/TT	−33.1	−92.9	−4.26
AT/TA	−30.1	−85.4	−3.67
TA/AT	−30.1	−89.1	−2.50
CA/GT	−35.6	−95.0	−6.12
GT/CA	−35.1	−93.7	−6.09
CT/GA	−32.6	−87.9	−5.40
GA/CT	−34.3	−92.9	−5.51
CG/GC	−44.4	−113.8	−9.07
GC/CG	−41.0	−102.1	−9.36
GG/CC	−33.5	−83.3	−7.66
Terminal A-T base pair	9.6	17.2	4.31
Terminal G-C base pair	0.4	−11.7	4.05

Reprinted with the permission from Ref. [9]. Copyright 1998, National Academy of Sciences, USA

Values of ΔH° and ΔS° have been predicted by nearest-neighbor method for all the ten possible pairs of nearest-neighbor interactions, as given in Table 2.1, along with the values of ΔG°_{37}. The parameters associated with the ten groups of neighbors shown in Table 2.1 are determined from melting points of short oligonucleotide duplexes [9]. The parameters calculated with the nearest-neighbor method agree well with experimental results.

Besides the stem structures, the loop structure, including loop sequences and loop length, also plays an important role in the stability of hairpins. Traditionally, the stem of a nucleic acid hairpin is regarded as a highly structured helix and the loop as a disordered coil. The thermodynamic stability of hairpins was attributed to the stem sequence. Single-stranded loops were assumed to destabilize the folded hairpin structure in an entropic manner based on the length of loops [11]. Thus, DNA hairpins with loop length of four to five residues were found to have maximum stability [12, 13], whereas early studies showed that RNA hairpin loops with six to seven nucleotides have highest stability [14]. These results correlated well with a simple structural principle [13]: that the steric hindrance effect for the folding of hairpin is mainly caused by the fact that the loop bridges the gap between the two complementary opposite sides of the stem. The enthalpy of DNA hairpin formation reaches a minimal value for loops of four to five nucleotides. On the other hand, the melting temperature decreases at increasing loop size based on the unfavorable entropy effect of loop formation [13]. Loop sequence has less effect on loop stability than its length. For DNA hairpins sharing the same stem, but with different 4-base loops, the order of stability was T loop > C loop > G loop > A loop, and the largest difference in melting temperature was 5–6 °C between the most stable T loop and least stable A loop [15].

In addition to the effect of stem and loop structures on the stability of hairpins, metal ions, such as Na^+ and Mg^{2+} ions, are also essential in stabilizing the folded hairpins through electrostatic interactions. Salt-dependent correction terms for the thermodynamic parameters of nucleic acid hybridization have been obtained from both experimental data [16, 17] and theoretical modeling [18, 19]. The salt dependence of hairpin stability has also been addressed [20].

2.2.3 Kinetics of Conformational Fluctuations of Hairpins

Kinetics studies have been conducted on the process of double-helix formation of RNA and DNA [21, 22]. Because of the fast kinetics of helix formation, a rapid kinetics approach is required. Using a laser or a capacitor to rapidly heat the water, the temperature jump (T-jump) method has been widely employed [23–25]. Controlled by a thermostatted bath, temperature jump is a perturbation method in which the system starts at equilibrium at a certain temperature (T_{init}) and ends at equilibrium at a higher temperature (T_{final}) [26]. Methods of duplex detection include absorbance, as well as fluorescence, optical activity, scattering, and conductivity. The relaxation of the system (τ) going from T_{init} to T_{final} is observed. In the open-to-closed transition of hairpin,

$$\text{coil} \underset{k_{-1}}{\overset{k_1}{\rightleftarrows}} \text{hairpin}$$

the observed relaxation rate, $1/\tau$, is given by the equation [3]

$$\frac{1}{\tau} = k_1 + k_{-1} \tag{2.4}$$

To extract k_1 and k_{-1} from τ, the equilibrium constant for the helix–coil transition, it is necessary to determine K by the optical or melting method. The relationship is given by:

$$K = \frac{k_1}{k_{-1}} \tag{2.5}$$

Combining Eqs. (2.4) and (2.5), both closing and opening rate constants can be solved. By study of fluorescence quenching using fluorescence correlation spectroscopy (FCS), it was found that closing rates depend on the sequence and length of the loop [3]. The results were confirmed by the laser T-jump method [23] and agreed well with a semiflexible polymer model [25]. A deviation from Arrhenius kinetics, i.e., the rate of a chemical reaction, was found in studies of Förster resonance energy transfer (FRET) fluctuations in DNA hairpins [27, 28]. Here, while the opening rate depends on the unzipping energy of the hairpin, it is essentially independent of loop characteristics, irrespective of loop size, sequence, and NaCl concentration [3].

2.3 Hybridization Thermodynamics and Kinetics of Molecular Beacons

2.3.1 Thermodynamics

Hybridization assays with sequence-specific oligonucleotide probes are commonly used techniques for the identification of complementary strands and the detection of genetic mutations and polymorphisms. However, linear oligonucleotide probes have limited abilities in cases where detection of single nucleotide polymorphism (SNP) is required. Specifically, an energy penalty is applied for any mismatched base pairs which destabilize the double helix. However, if the duplex is sufficiently long, the free energy penalty from a single-base mismatch becomes negligible [29]. Therefore, the mismatched base pair only minimally affects the stability of the resulting duplex. It is often necessary to use proteins or other chemicals to assist in the identification of mismatched base pairs [30, 31].

During the last two decades, the emergence of molecular beacons, which are stem–loop probes, has greatly improved our capabilities in the analytical, biochemical, and biomedical fields. Compared with linear nucleic acid probes, one of the most attractive features of molecular beacons is enhanced specificity by the ability to form a stem–loop structure. The hairpin structure enables molecular beacons to discriminate between targets with even a single nucleotide. The competition between the formation of unimolecular hairpin and the probe–target hybridization reduces the probability of mismatched hybridization events from the relatively low stability [32–34].

2.3.1.1 Determination of Thermodynamic Parameters

As shown in Fig. 2.2, linear nucleic acid probes have two possible states with their targets: linear probe–target duplex and random coil. In contrast, molecular beacons in solution with their targets can have at least three distinct states: molecular beacon–target duplex (phase 1), stem–loop hairpin (phase 2), and random coil (phase 3) [2, 35].

Following Bonnet et al. [36] and Tsourkas et al. [2], the dissociation constant $K_{2\text{-}3}$ corresponding to the transition between stem–loop hairpin and random coil is given by:

$$K_{2-3}(\theta) = \left(\frac{F-\beta}{\gamma-F}\right) \tag{2.6}$$

where F is the fluorescence intensity at a given temperature θ, β is the fluorescence of molecular beacons in the hairpin form (obtained at low temperatures, such as 10 °C), and γ is the fluorescence of molecular beacons in the random-coil form (obtained at high temperatures, such as 80 °C) in the absence of targets [36]. Based

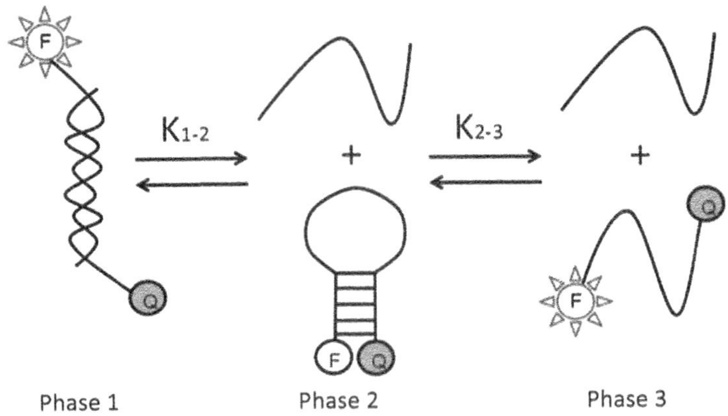

Fig. 2.2 Phase transitions of molecular beacons with targets in solution. Phase 1: fluorescent molecular beacon–target duplex; phase 2: nonfluorescent stem–loop hairpin; and phase 3: fluorescent random coil

on the Gibbs relationship,

$$\Delta G^{\circ}_{2-3} = \Delta H^{\circ}_{2-3} - \theta \Delta S^{\circ}_{2-3} \tag{2.7}$$

and the relationship between free energy and equilibrium constant,

$$\Delta G^{\circ}_{2-3} = -RT\,ln\,K_{2-3}(\theta) \tag{2.8}$$

fluorescence and temperature can be correlated by

$$R\,ln\left(\frac{F-\beta}{\gamma-F}\right) = -\Delta H^{\circ}_{2-3}\frac{1}{\theta} + \Delta S^{\circ}_{2-3} \tag{2.9}$$

where R is the gas constant (1.9872 cal mol^{-1} K^{-1}), T is the temperature in Kelvin, and $\Delta H_{2\text{-}3}$ and $\Delta S_{2\text{-}3}$ are the changes in enthalpy and entropy in the transition, respectively, which are determined by fitting the fluorescence–temperature data into a straight line using this equation.

The thermodynamic parameters for the transition from a molecular–target duplex (phase 1) to a stem–loop hairpin (phase 2), ΔH°_{1-2} and ΔS°_{1-2}, were determined by fluorescence measurements of the thermal denaturation profiles in the presence of targets at different concentrations. At the melting temperature of the duplex (θ_m), the concentration of molecular beacons in the hairpin form (M_{hairpin}) equals the concentration of the probe–target duplex (MT). Therefore, based on the equation

$$K_{1-2} = \frac{[M_{\text{hairpin}}][T]}{[MT]} \tag{2.10}$$

$K_{1\text{-}2} = T_m = T_0 - 0.5M_0$, where T_m is the concentration of free target and T_0 and M_0 are the initial concentrations of targets and molecular beacons, respectively. The thermodynamic parameters ΔH°_{1-2} and ΔS°_{1-2} can be determined by fitting the concentration-melting temperature data into a straight line using this equation, which establishes the relationship between melting temperature and the thermodynamic parameters, ΔH°_{1-2} and ΔS°_{1-2}:

$$RlnK_{1-2}(\theta_m) = Rln(T_0 - 0.5M_0) = -\Delta H^{\circ}_{1-2}\frac{1}{\theta_m} + \Delta S^{\circ}_{1-2} \tag{2.11}$$

The dissociation constant $K_{1\text{-}2}$ characterizing the transition between molecular beacon–target duplex and the stem–loop hairpin can be determined from the fluorescence data obtained by the melting measurements of molecular beacons in the presence of target [36], by the equation

$$K_{1-2}(T) = \frac{(\alpha - F)T_0}{(F - \beta) + (F - \gamma)K_{2-3}} \tag{2.12}$$

where α is the fluorescence intensity of molecular beacon–target duplex at low temperatures such as 10 °C, and F and β are the same as above mentioned in Eq. (2.6).

Equilibrium states of molecular beacons in the presence of targets can be significantly influenced by the probe (loop) length. For example, Tsourkas et al. found that both ΔH°_{1-2} and ΔS°_{1-2} increase when probe length increases from 17 to 19 bases for a molecular beacon with a five-base stem. It was also found that the stem length of a molecular beacon had greater effect on its equilibrium state than the loop length [2, 37]. The effect of mismatched base pairs on the equilibrium states of molecular beacons was also studied [2, 36]. The target with single-base mismatches showed a less favorable binding with molecular beacons. Furthermore, a centrally positioned mismatch had a greater impact on equilibrium than a terminal mismatch in the probe domain of a molecular beacon [2, 36].

2.3.1.2 Melting Temperature

The stability of probe–target duplex is expressed in terms of melting temperature. To demonstrate the effect of molecular beacon structure on melting behavior of the probe–target duplex, the melting temperature was studied by varying the loop and stem structures. Tsourkas et al. found that melting temperature increased with probe length. Stem length also had the same effect on melting temperature [2]. For three molecular beacons, each having 17 bases in the loop domain, the melting temperature decreased by 7.8 °C when the stem length was increased from four to six. Moreover, when probe length was shorter, stem length was observed to have a relatively greater impact on melting temperature.

Table 2.2 Melting temperatures for the dissociation of a perfectly complementary probe–target duplex (first entry), probe–target duplexes containing different mismatched base pairs at the same position (next three entries), and probe–target duplexes containing the same mismatched base pair at different positions (last nine entries)

Mismatch	Position (n)	θ_m (°C)
T-A	0	42
A-A	0	27
C-A	0	23
G-A	0	28
G-A	−4	30
G-A	−3	29
G-A	−2	30
G-A	−1	29
G-A	0	28
G-A	+1	29
G-A	+2	29
G-A	+3	29
G-A	+4	31

The presence of mismatched bases also has a significant effect on the melting temperature of molecular beacon–target duplex. As shown by Bonnet et al. in Table 2.2, single-base mismatches greatly reduce the melting temperature of the probe–target duplex by more than 10 °C [36]. However, the position of the single-base mismatches did not have any impact on the stability of the probe–target duplex, which was consistent with the prediction made using an all-or-none mechanism whereby an action potential occurs fully or not at all [38]. The identity of the mismatched base pairs also had no appreciable impact on the melting temperature.

The melting temperature of probe–target duplex may also be influenced by the fluorophore–quencher pairs. It is well known that most fluorophores and quenchers used to label molecular beacons contain hydrophobic functional groups. When labeled at the ends of molecular beacons, fluorophore–quencher pairs can have hydrophobic interactions if they are in the close proximity. Marras et al. reported an increase of 3 °C in the melting temperature of linear probe–target duplex for Fluorescein–Dabcyl pair and 5 °C for Cy3–Dabcyl pair [39]. A similar effect was observed by Tsourkas et al. [2]. Compared with the values predicted using nearest-neighbor methods, as defined above, an increase of the observed melting temperature was reported.

2.3.1.3 Specificity of Molecular Beacons

Generally, probe specificity is expressed by the difference in melting temperature $\Delta\theta_m$ between the duplex of probe-perfect target and the duplex of probe-mismatched target. For molecular beacons, it has been found that longer stem length results in an improved ability to distinguish between wild-type and mutant targets over a broader range of temperatures [2, 37]. Molecular beacons with longer stems have better stability in the stem–loop conformation and smaller difference in free energy between closed state and molecular beacon–target duplexes. This generates a reduced binding affinity between single-base mismatched targets and molecular beacons. Tsourkas et al. calculated the fraction of molecular beacon–target duplex at different temperatures for molecular beacons with stems of different lengths and found that $\Delta\theta_m$ increased with the stem length, as shown in Fig. 2.3a [2]. To further evaluate stem length relative to the specificity of molecular beacons in discriminating wild-type and mutant targets, the difference in fraction of probe–target duplex is analyzed. As reported by Tsourkas et al. and shown in Fig. 2.3b below, molecular beacons with 6-base stems were able to discriminate targets over a broader range of temperatures. It was also demonstrated by Tsourkas et al. that molecular beacons with longer probe length had lower specificity.

Specificity and affinity are two major factors that determine the efficiency of nucleic acid probes. In general, sequence specificity and binding affinity of nucleic acid probes are negatively correlated with each other [40, 41]. Figure 2.4 shows the fraction of correct and mismatched complexes for different binding affinities of an oligonucleotide probe. In the graph, a window can be determined where both high affinity and high specificity can be realized. Beyond this window, the specificity decreases, as binding affinity increases. However, because the window of optimal conditions is quite narrow for linear probes, it is highly desirable to find ways to enlarge the optimal range.

Compared with linear probes, one of the most significant advantages of molecular beacons is the higher specificity with which they recognize target sequences [2, 36, 37, 42–44]. The enhanced specificity of molecular beacons is attributed to the three-phase transition of molecular beacons upon interaction with targets, whereas linear nucleic acid probes have only two possible phases: probe–target duplex and random coil. Molecular beacons are capable of forming a hairpin structure that is weaker than the duplex with a correct target, but stronger than the duplex with a mismatched target, as shown in Fig. 2.5 [40, 45, 46].

The enhanced specificity of molecular beacons compared with linear probes can be evaluated by the difference in the complex fraction between probe-wild-type target and probe-mutated target, $\Delta\alpha = \alpha_{WT} - \alpha_{target\ B}$ [2, 47]. Compared with linear probes, all molecular beacons with different stem and probe lengths demonstrate an improved ability to discriminate mismatched target (Fig. 2.6). Molecular beacons can not only discriminate between wild-type and mutant targets over a broader range of temperatures, but also maintain a larger difference in fraction of bound molecular beacons between wild-type and mutant targets.

The free energy diagram was also employed to explain the enhanced specificity of molecular beacons [36, 40]. Figure 2.7 shows how molecular beacons can widen the range of optimal conditions for affinity and specificity. In the free energy diagram plotted as a function of temperature, linear probes have an optimal temperature range of $\Delta\theta$, whereas molecular beacons have a much wider range of $\Delta\theta'$. The range is also shifted to lower temperatures where the correct probe–target duplex is more stable. Since molecular beacons are constrained polymers

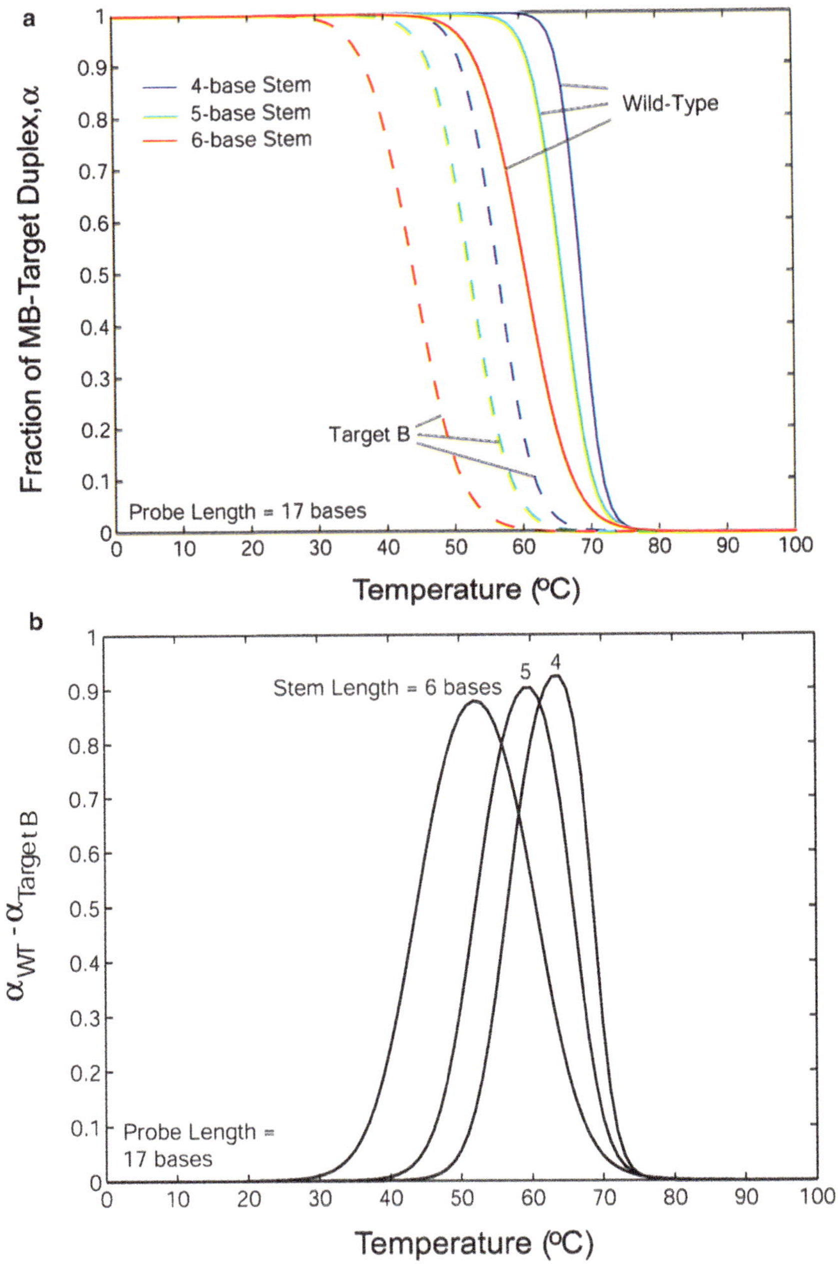

Fig. 2.3 (**a**) Melting curves for molecular beacons hybridizing to wild-type target (*solid line*) and target B (*dashed line*) and (**b**) the difference in the fraction of molecular beacons bound to wild-type target and the fraction of MBs bound to mutant target B. Molecular beacons have a probe length of 17 bases and stem lengths of 4, 5, and 6 bases (Reprinted with the permission from Ref. [2]. Copyright 2003, Oxford University Press, Inc.)

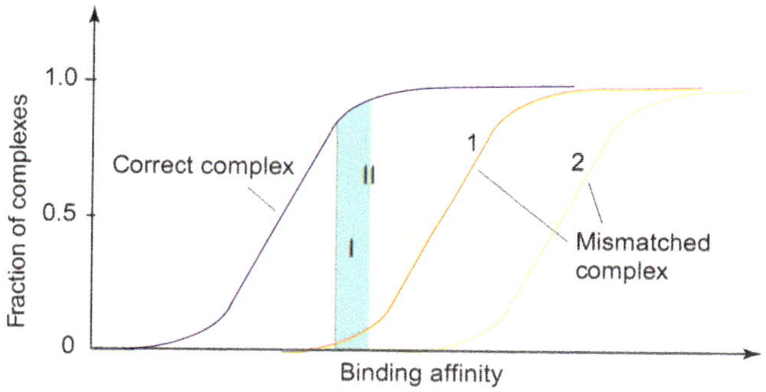

Fig. 2.4 Equilibrium binding of an oligonucleotide probe to correct and mismatched DNA or RNA targets as a function of their binding affinity (Reprinted from Ref. [40], Copyright 2004, with permission from Elsevier)

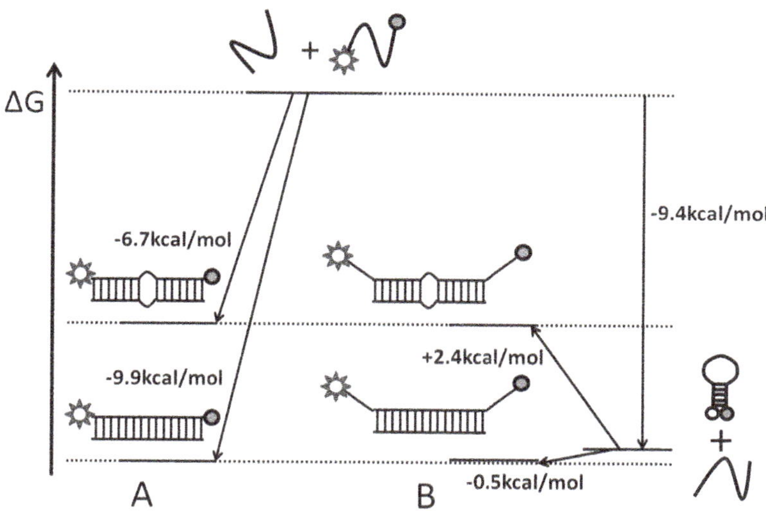

Fig. 2.5 Free energy diagram of hybridization. (**a**) A classical denatured probe of 15 bases; the *two arrows* correspond to the hybridization with a perfect matched target or a one-base-mismatched target. In both cases, complete hybridization takes place, as the free energies are negative. (**b**) A molecular beacon with a loop of 15 bases and a stem of 13 base pairs; in this case, the beacon hybridizes to its perfect matched target (negative free energy), but not to the one-base-mismatched target. The free energy values are calculated according to Eqs. (2.7) and (2.8) found in Sect. 2.3. A schematic of the molecular beacon is shown on the *right* (Reprinted from Ref. [45], Copyright 1999, with permission from Elsevier)

that undergo a greater reorganization than unstructured probes upon formation of probe–target duplexes, they have a better ability to sense a mismatch. Similarly, all conformationally constrained probes should display higher specificity in molecular recognition than unstructured probes.

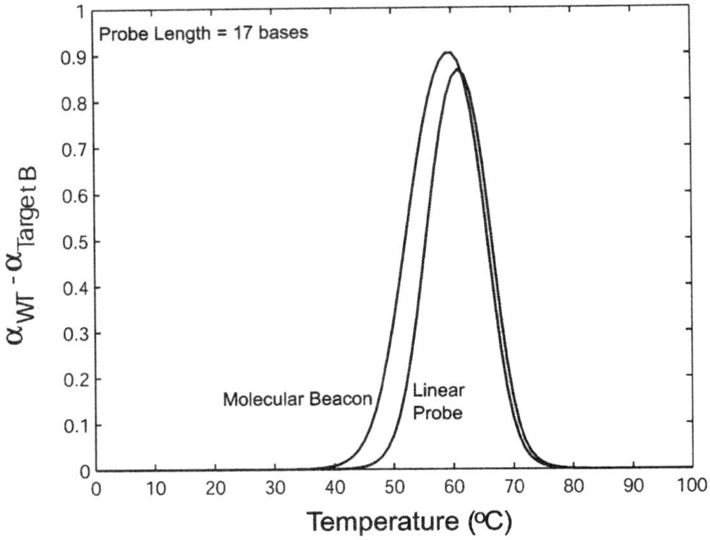

Fig. 2.6 The fraction of molecular beacons compared to the fraction of linear probes bound to wild-type target and mutant target B (Reprinted with the permission from Ref. [2]. Copyright 2003, Oxford University Press, Inc.)

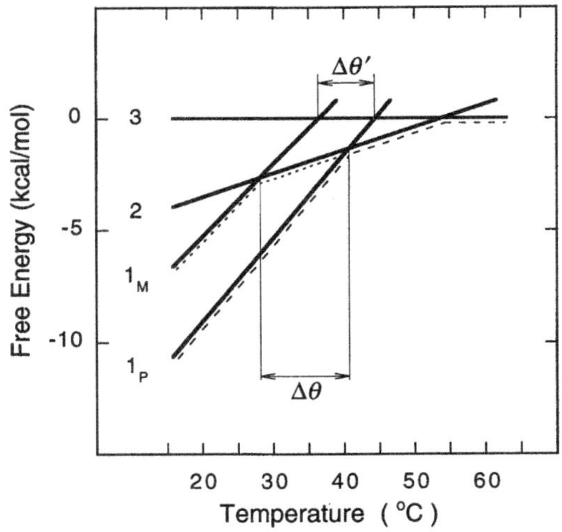

Fig. 2.7 Free energy of the three phases of a solution of molecular beacons in equilibrium with target oligonucleotides. The equation of each line is $\Delta G = \Delta H - \theta \Delta S$. *Straight lines* indicate the free energy of a molecular beacon (curve 2), a linear probe (curve 3), and its duplex complexes with correct (curve 1p) and mismatched (curve 1M) targets (Reprinted with the permission from Ref. [36]. Copyright 1999, National Academy of Sciences, USA)

2.3.1.4 Dynamic Range and Detection Limit

Upon target binding, the three-phase transition of molecular beacons yields different signaling states, and the nonbinding, nonsignaling state is shifted toward the binding, signaling state. Although a larger signal change is induced when the equilibrium constant is shifted toward the nonbinding state because of the lower background, such signal also reduces the affinity because of a larger conversion of free energy. Thus, similar to other biomolecular switches, the MB's dynamic range and detection limit can be tuned by changing its thermodynamics [48, 49].

As reported by Vallée-Bélisle et al. the three-phase transition model is used to predict the relationship between switching thermodynamics and observed affinities K_D^{obs} by the equation below:

$$K_D^{\mathrm{obs}} = K_D^{\mathrm{int}} \left(\frac{1 + K_\mathrm{s}}{K_\mathrm{s}} \right) \tag{2.13}$$

where K_D^{int} is the intrinsic affinity of the molecular beacon–target duplex and K_s is the switching equilibrium constant from closed hairpin state to random coil [50]. It was found that shifting the equilibrium constant toward the nonbinding state by ten times also increased the dynamic range by ten times which could be done by substituting one A-T with G-C pair in the stem. The upper limit of the dynamic range of concentration could be increased by shifting the equilibrium constant toward the nonbinding state. However, the lowest limits (detection limit) are reached at intermediate values of K_S.

2.3.2 *Kinetics of Molecular Beacons*

The structure of molecular beacons significantly impacts the thermodynamic properties of the molecular beacon–target system, including melting temperature and specificity. Similarly, the kinetics is also influenced by the structure of molecular beacons. For example, longer stems result in slower hybridization kinetics of molecular beacons with target, while molecular beacons with longer loops show larger kinetic constants.

Kinetics information can be obtained by fluorescence measurements [2, 37, 51, 52]. In the analysis of the hybridization kinetics, the association of two oligonucleotides is a second-order reaction, whereas the dissociation is first order, and the reaction can be expressed by the following equation [2]:

$$B + T \underset{k_2}{\overset{k_1}{\rightleftarrows}} D$$

$$\frac{\mathrm{d}[D]}{\mathrm{d}t} = k_1[B][T] - k_2[D] \tag{2.14}$$

where $[B]$, $[T]$, and $[D]$ are the concentrations of molecular beacon, target, and molecular beacon–target complex, respectively, and k_1 and k_2 are the rate constants of formation and dissociation of the molecular beacon–target complex, respectively. The solution of the equation is

$$1-\frac{[D(t)]}{[D_{eq}]}=e^{-\Delta k_1 t}\left[1-\lambda\frac{[D(t)]}{[D_{eq}]}\right] \quad (2.15)$$

where $\Delta = \sqrt{(B_0+T_0+K_{12})^2-4B_0T_0}$, $[D_{eq}] = \frac{1}{2}(B_0+T_0+K_{12}-\Delta)$, $\lambda = [D_{eq}]^2/B_0T_0$, and $K_{12}=k_2/k_1$ is the dissociation constant of the molecular beacon–target duplex. Based on the all-or-nothing assumption, fluorescence intensity of the molecular beacon–target system is proportional to the concentration of molecular beacon–target complex. Thus it can be assumed that

$$\frac{[D(t)]}{[D_{eq}]}=\frac{F(t)-F_0}{F_{eq}-F_0} \quad (2.16)$$

where $F(t)$ is the fluorescence intensity at time t, F_0 is the initial fluorescence intensity, and F_{eq} is the fluorescence intensity at equilibrium. The rate constants k_1 and, subsequently, $k_2=K_{12}k_1$ were obtained using two methods. Two different curve-fitting methods can be used to obtain the on-rate k_1 and off-rate k_2. One method uses a nonlinear least-square method to fit fluorescence data directly to Eq. (2.15). The second method fits the data to a logarithmic form of Eq. (2.15) with a straight line (the slope is k_1), as shown in this equation:

$$\frac{1}{\Delta}\ln\left(1-\frac{F(t)-F_0}{F_{eq}-F_0}\right)=\frac{1}{\Delta}\ln\left(1-\lambda\frac{F(t)-F_0}{F_{eq}-F_0}\right)-k_1 t \quad (2.17)$$

In the example given by Tsourkas et al. the normalized fluorescence restoration was monitored as a function of time after introduction of targets with 17 bases in the loop and four, five, and six bases in the stems. The on-rate constants of molecular beacons with stems and loops of different lengths are plotted and compared. Results showed that molecular beacons always had smaller on-rate constants and that a decrease of stem length significantly increased the on-rate constants. For molecular beacons with 17 bases in the loop, the on-rate constant decreased tenfold when the stem length increased from 5 to 6 bases. Moreover, the probe length can also influence the on-rate constant. Generally, longer probes had faster hybridization kinetics, and this effect was greater for molecular beacons with longer stems.

Different methods have been used to investigate the kinetic properties of DNA hybridization by measuring comparable hybridization/dehybridization rate constants over a wide range of temperature. Based on the results obtained by Chen et al. and Tsourkas et al. a three-step mechanism was proposed to explain the kinetics of hybridization between molecular beacons and targets, as shown in Fig. 2.8b [2, 53, 54]. In the first step, similar with the case of linear probes, the target gets into the close proximity of the probe domain of the hairpin molecular beacon through formation of a few base pairs into a transient intermediate called a nucleus. In the second step, the stem opens along with the formation of more base pairs

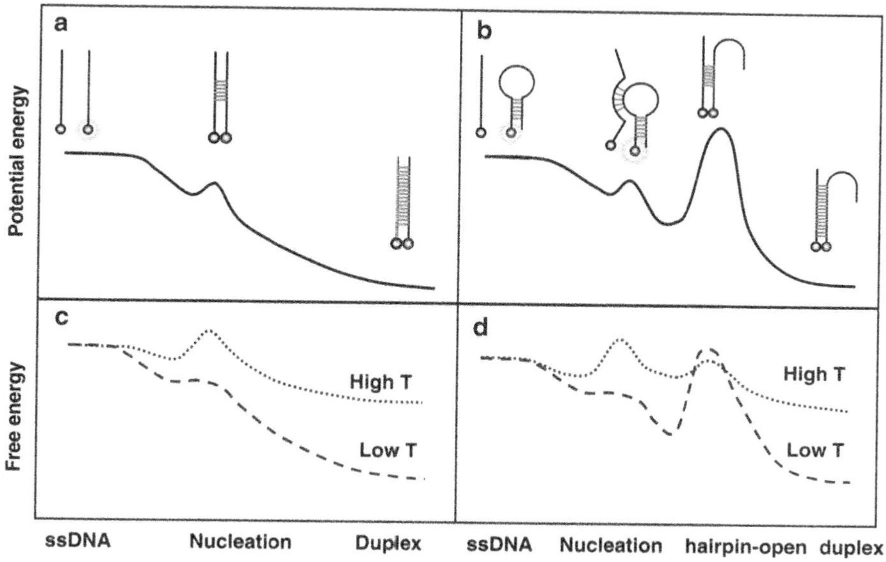

Fig. 2.8 Schematic view of the profiles of potential energy (*solid line*) and free energy at high temperature (*dotted line*) and low temperature (*dashed line*), as a function of reaction progress coordinate for hybridization of random-coil DNA (**a** and **c**) and molecular beacon (**b** and **d**) (Reprinted with the permission from Ref. [54]. Copyright 2007, Oxford University Press, Inc.)

between molecular beacon and target. In the third step, the complete hybridization is achieved after the formation of the remaining base pairs. The rate-limiting step is the first and second step, while the third step is much faster.

Based on standard chemical kinetic theory, a metastable intermediate exists in the rate-limiting step indicated by a negative activation energy for linear probes as shown in Fig. 2.8a [54]. Furthermore, as shown in Fig. 2.8c, at high temperatures, the rate-limiting step is the nucleation of the first few base pairs represented by a negative activation energy. However, at low temperature, the rate-limiting step switches to the diffusion controlled reaction [55, 56]. Compared with the hybridization of linear DNA probes, the hybridization of molecular beacon with target has to overcome one more energy barrier represented by the opening of the hairpin stem (Fig. 2.8b). This free energy barrier decreases as the temperature increases. Therefore, the rate-limiting step switches from the opening of stem to nucleation at high temperatures, as shown in Fig. 2.8d.

In addition, molecular beacons with short stems possess faster binding kinetics with target, while longer stems give a higher energy barrier, resulting in a smaller rate constant in the hybridization process. Molecular beacons with long stems and short probes also show on-rate constants which depend on the probe length because the energy gain in the hybridization of probes barely compensates for the cost for opening the stems.

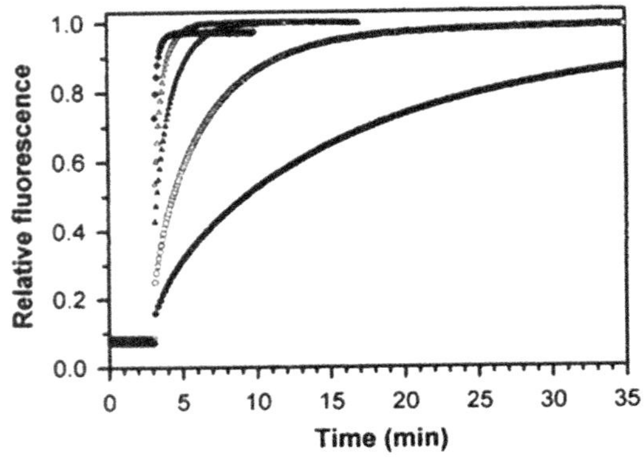

Fig. 2.9 Kinetic responses of molecular beacons upon hybridization to targets at 25 °C in 20 mM Tris–HCl buffer (pH 8.0) containing MgC12 at different concentrations: 1 (●), 2.5 (○), 5(▲), 10 (Δ), and 50 mM (♦) (Reprinted with the permission from Ref. [42]. Copyright 2002 American Chemical Society)

2.3.3 *Effect of Buffer Conditions on Molecular Beacons*

The majority of studies reporting on DNA hybridization were performed in water. Buffer composition has a profound impact on probe performance. Since DNA is a polyanion, it has been reported that a faster hybridization rate could be obtained by increasing the salt concentration [55, 56]. Studies have also reported on DNA hybridization in different organic solvents. McConaughy et al. reported that formamide could accelerate DNA hybridization [57]. It was also reported by Kohne et al. that a water-phenol two-phase system increased the hybridization rate by interfacial adsorption and diffusion [58].

Salt concentration also has an important effect on the hybridization of molecular beacons with targets. Kuhn et al. demonstrated that molecular beacons respond differently to different salt concentrations [42]. Figure 2.9 shows the normalized kinetic responses of molecular beacon–target hybridization at different concentrations of $MgCl_2$. It can be seen that molecular beacons hybridize faster with their targets at higher salt concentrations. Yao et al. [59] reported on the use of immobilized molecular beacons for DNA analysis and observed no fluorescence change in the absence of $MgCl_2$, suggesting that $MgCl_2$ is required for the hybridization of the MB with its target DNA. The fluorescence intensity of an immobilized MB was found to decrease with the increase of $MgCl_2$ concentration from 3 to 200 mM. However, while the fluorescence of the MB–cDNA was enhanced when the $MgCl_2$ concentration was increased from 0 to 100 mM, the fluorescence intensity of MB–cDNA was decreased when the $MgCl_2$ concentration was further increased to 200 mM.

Systematic studies of the effects of different solvents on molecular beacon performance were also carried out. Dave et al. demonstrated DNA detection in nine different organic solvents, each varying up to 75 % (v/v). Compared with DNA detection in water, there are several important features for detection in organic solvents. First, hybridization of the molecular beacon with its target DNA happened in solvent systems containing all nine solvents up to a certain percentage. Second, the hybridization kinetics in most organic solvents was significantly faster than that in water. For example, as shown in Fig. 2.10, the hybridization rate of molecular

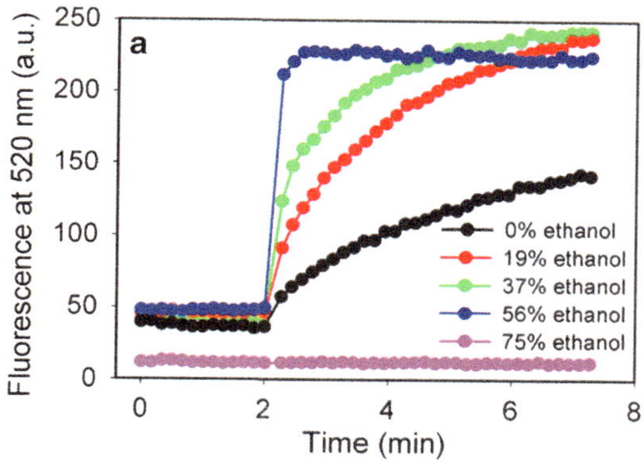

Fig. 2.10 Kinetics of molecular beacon hybridization in 300 mM NaCl and 10 mM HEPES, pH 7.6, with different ethanol percentages (Reprinted with the permission from Ref. [91]. Copyright 2010 American Chemical Society)

beacon with target was enhanced by 70 times in solution with 56 % ethanol. Third, the ability to discriminate single-base mismatch was still maintained in the organic solvent for molecular beacons. Finally, the melting temperature of the molecular beacon–target duplex decreased as the percentage of organic solvents increased.

2.3.4 Optimization of Selectivity and Kinetics of Molecular Beacons

Excellent selectivity is a direct result of hairpin structure and stands out as a major advantage compared to linear probes. Selectivity can be readily improved by optimizing the structure of molecular beacons [2, 44]. From the thermodynamic point of view, the simplest way to improve specificity is to increase the number of base pairs or the G-C content in the stem region. However, although a more stable stem is beneficial to the specificity of molecular beacons, it hinders hybridization kinetics [2]. On the other hand, molecular beacons with short stems have faster hybridization kinetics and improved target affinities, but suffer from a lower signal-to-background ratio and selectivity. The impact of probe length on the behavior of molecular beacons is less significant than that of stem length. Therefore, the selection of the optimal molecular beacon for a specific application appears to be difficult, and a compromise must be found between higher selectivity and faster hybridization. In general, the stem contains about five to seven base pairs, while the loop has 15–25 bases.

Both selectivity and hybridization kinetics are significantly affected by temperature. Therefore, temperature should be considered as an important external factor for designing molecular beacons for certain applications. Higher temperatures drive the opening of molecular beacons to form random coils, generating high background. In order to improve the performance of molecular beacons, especially for applications with relatively high temperatures, alternative strategies have also been developed. First, different from the traditional design of using target-irrelevant stem, a shared-stem design has been utilized [37]. For this type of molecular beacon, one arm of the stem participates in both stem formation of the closed hairpin and the target hybridization of the opened conformation. This design should be able to improve FRET efficiency based on restricted movement of the dye linked with the shared arm [60]. It also results in higher melting temperatures, as indicated by the thermodynamic studies [37].

Second, different types of synthetic nucleic acids were used to create molecular beacons. For live-cell RNA imaging applications, phosphorothioate DNA [61], 2′-*O*-methyl RNA [47, 62], 2′-*O*-methyl RNA/DNA chimeras [63], peptide nucleic acid (PNA) [42], and locked nucleic acid (LNA) [64, 65] have been employed to construct molecular beacon probes. Nucleic acid affinity for RNA is ranked from highest to lowest: LNA, PNA, 2′-*O*-methyl RNA, RNA, and DNA [66–68]. At first glance, high-affinity probes, by their high melting temperatures, may result in a corresponding high degree of nonspecific hybridization at 37 °C. However, compared with studies performed in solution with high free ion concentrations, studies carried out with fewer ions in the intracellular environment [69, 70] may result in melting temperatures with less effect on hybridization. Therefore, specificity could still be maintained for high-affinity probes in live cells. Tsourkas et al. performed a detailed study of 2′-*O*-methyl and 2′-deoxy molecular beacons in the presence of RNA and DNA targets and found improved stability for 2′-*O*-methyl/RNA duplex accompanied with faster hybridization kinetics [51]. Since LNA–DNA hybridization is stronger than DNA–DNA hybridization, it has also been reported that LNA-molecular beacons are more selective than DNA molecular beacons [71, 72]. As reported by Wang et al. the LNA-MBs remained in a hairpin structure even at 95 °C and had a capability for single nucleotide polymorphism (SNP) detection superior to that of DNA molecular beacons [73].

2.3.5 Surface-Immobilized Molecular Beacons

Microarray technology has made profound contributions to the field of molecular biology with various applications, including pathogen detection, disease diagnostics, gene expression profiling, and drug discovery [74–77]. Surface immobilization of DNA probes allows spatially multiplexed detection, which dramatically accelerates many types of investigation. Compared with linear nucleic acid probes, surface-immobilized molecular beacons possess better specificity and higher signal-to-background ratio.

Comparative studies of DNA hybridization in solution and on solid-solution interface have been performed to analyze the differences in thermodynamic and kinetic properties [78, 79]. It was revealed that hybridization on the solid-solution interface was significantly slower than solution phase and that the hairpin structures in molecular beacon probes further slowed down the hybridization kinetics. Moreover, the traditional model describing the hybridization process in solution phase was not applicable on the solid-solution interface.

In fact, the hybridization of nucleic acid probes with their targets is significantly affected by many factors, such as surface strand density [80], surface charge [81], brush effect [82], point mismatch [83], DNA length [84], and flatness of the substrate [85]. These factors make it necessary to consider the surface interactions between molecular beacons and their targets which are different from those in solution, such as interfacial concentration gradient, probe–interface interactions, and steric hindrance. For example, surface electrostatic interactions greatly affect the binding of surface-immobilized probes with their targets. Nucleic acid targets can be repelled or attracted to the surface depending on the charge of the surface material. Electrostatic repulsion between nucleic acid targets and the surface can block hybridization events. Similarly, steric hindrance caused by increasing probe densities on the surface can further reduce the hybridization efficiency of surface-immobilized probes from repulsive electrostatic interactions.

In general, surface-immobilized molecular beacon probes exhibit lower sensitivities compared with those in solution. The lower sensitivity is partially attributed to the inefficient quenching of the molecular beacons on the surface caused by their nonspecific interactions with the supporting materials. These interactions tend to destabilize the hairpin structures of molecular beacons. Glass has been widely used as a solid support for the immobilization of MBs [59, 86]. Unfortunately, glass suffers from interfacial effect of static charging which partially opens molecular beacons, resulting in high background signals for the immobilized molecular beacons in the absence of targets. Molecular beacons have also been immobilized on porous surfaces using hydrogel films, which create a solution-like environment [87]. However, the gel network slows down the hybridization process by restricted transport of targets. Gold surfaces have also been employed as substrates for immobilization of molecular beacons as a consequence of the easy chemistry linkage, as well as the fluorescence quenching ability, of gold [88, 89]. However, nitrogen-based functional groups on DNA bases can be nonspecifically adsorbed to the gold surfaces through chemical adsorption [88]. Another issue concerning molecular beacons immobilized on gold surfaces is the nonuniform distribution of probes which affects sensitivity, specificity, and kinetics [88, 90].

2.4 Summary

Molecular beacon probes with high specificity and sensitivity have become an important tool in biomedical and bioanalytical studies. The advantages of molecular beacons are attributed to the flexible stem–loop hairpin structures. The conformational fluctuations of molecular beacons in the absence of targets can be described using the equilibrium between the open and closed states. In the presence of targets, the interaction between molecular beacons and targets is characterized by a three-phase model: hybridized with a target, free in the stem–loop conformation, and free as a random coil. The thermodynamics and kinetics of molecular beacons are mainly determined by their hairpin structures and sequences, but can be equally affected by many other factors, such as temperature, salt concentration, solvents, and probe–surface interactions. To optimize the performance of molecular beacons, factors that affect affinity, specificity, and kinetics must be considered and a probe design and experimental conditions carefully chosen on the basis of the purpose of different applications.

References

1. Tyagi S, Kramer FR (1996) Molecular beacons: probes that fluoresce upon hybridization. Nat Biotechnol 14:303–308
2. Tsourkas A, Behlke MA, Rose SD, Bao G (2003) Hybridization kinetics and thermodynamics of molecular beacons. Nucleic Acids Res 31:1319–1330
3. Bonnet G, Krichevsky O, Libchaber A (1998) Kinetics of conformational fluctuations in DNA hairpin-loops. Proc Natl Acad Sci U S A 95:8602–8606
4. Vallone PM, Benight AS (1999) Melting studies of short DNA hairpins containing the universal base 5-nitroindole. Nucleic Acids Res 27:3589–3596
5. Antao VP, Tinoco I (1992) Thermodynamic parameters for loop formation in RNA and DNA hairpin tetraloops. Nucleic Acids Res 20:819–824
6. Vallone PM, Paner TM, Hilario J, Lane MJ, Faldasz BD, Benight AS (1999) Melting studies of short DNA hairpins: influence of loop sequence and adjoining base pair identity on hairpin thermodynamic stability. Biopolymers 50:425–442
7. Rentzeperis D, Shikiya R, Maiti S, Ho J, Marky LA (2002) Folding of intramolecular DNA hairpin loops: enthalpy-entropy compensations and hydration contributions. J Phys Chem B 106:9945–9950
8. Breslauer KJ, Frank R, Blöcker H, Marky LA (1986) Predicting DNA duplex stability from the base sequence. Proc Natl Acad Sci U S A 83:3746–3750
9. SantaLucia J (1998) A unified view of polymer, dumbbell, and oligonucleotide DNA nearest-neighbor thermodynamics. Proc Natl Acad Sci U S A 95:1460–1465
10. Owczarzy R, Vallone PM, Gallo FJ, Paner TM, Lane MJ, Benight AS (1997) Predicting sequence-dependent melting stability of short duplex DNA oligomers. Biopolymers 44:217–239
11. Jaeger JA, Turner DH, Zuker M (1989) Improved predictions of secondary structures for RNA. Proc Natl Acad Sci U S A 86:7706–7710
12. Haasnoot C, De Bruin S, Berendsen R, Janssen H, Binnendijk T, Hilbers C, Van der Marel G, Van Boom J (1983) Structure, kinetics and thermodynamics of DNA hairpin fragments in solution. J Biomol Struct Dyn 1:115–129
13. Haasnoot CA, Hilbers CW, van der Marel GA, van Boom JH, Singh UC, Pattabiraman N, Kollman PA (1986) On loop folding in nucleic acid hairpin-type structures. J Biomol Struct Dyn 3:843–857
14. Tinoco I, Borer PN, Dengler B, Levine MD, Uhlenbeck OC, Crothers DM, Gralla J (1973) Improved estimation of secondary structure in ribonucleic acids. Nature 246:40–41
15. Senior MM, Jones RA, Breslauer KJ (1988) Influence of loop residues on the relative stabilities of DNA hairpin structures. Proc Natl Acad Sci U S A 85:6242–6246
16. Schildkraut C, Lifson S (1965) Dependence of the melting temperature of DNA on salt concentration. Biopolymers 3:195–208
17. Owczarzy R, You Y, Moreira BG, Manthey JA, Huang L, Behlke MA, Walder JA (2004) Effects of sodium ions on DNA duplex oligomers: improved predictions of melting temperatures. Biochemistry (NY) 43:3537–3554
18. Tan ZJ, Chen SJ (2006) Nucleic acid helix stability: effects of salt concentration, cation valence and size, and chain length. Biophys J 90:1175–1190

19. Tan ZJ, Chen SJ (2007) RNA helix stability in mixed Na/Mg_2 solution. Biophys J 92:3615–3632
20. Tan ZJ, Chen SJ (2008) Salt dependence of nucleic acid hairpin stability. Biophys J 95:738–752
21. Turner D, Sugimoto N, Freier S (1990) Thermodynamics and kinetics of base-pairing and of DNA and RNA self-assembly and helix coil transition. Nucleic Acids 1:201–227
22. Bloomfield VA, Crothers DM, Tinoco I (2000) Nucleic acids: structures, properties, and functions. University Science Books, Sausalito
23. Ansari A, Kuznetsov SV, Shen Y (2001) Configurational diffusion down a folding funnel describes the dynamics of DNA hairpins. Proc Natl Acad Sci U S A 98:7771–7776
24. Shen Y, Kuznetsov SV, Ansari A (2001) Loop dependence of the dynamics of DNA hairpins. J Phys Chem B 105:12202–12211
25. Kuznetsov SV, Shen Y, Benight AS, Ansari A (2001) A semiflexible polymer model applied to loop formation in DNA hairpins. Biophys J 81:2864–2875
26. Bevilacqua PC, Blose JM (2008) Structures, kinetics, thermodynamics, and biological functions of RNA hairpins. Annu Rev Phys Chem 59:79–103
27. Wallace MI, Ying L, Balasubramanian S, Klenerman D (2001) Non-Arrhenius kinetics for the loop closure of a DNA hairpin. Proc Natl Acad Sci U S A 98:5584–5589
28. Wallace MI, Ying L, Balasubramanian S, Klenerman D (2000) FRET fluctuation spectroscopy: exploring the conformational dynamics of a DNA hairpin loop. J Phys Chem B 104:11551–11555
29. Wemmer DE, Chou SH, Hare DR, Reid BR (1985) Duplex-hairpin transitions in DNA: NMR studies on CGCGTATACGCG. Nucleic Acids Res 13:3755–3772
30. Youil R, Kemper BW, Cotton R (1995) Screening for mutations by enzyme mismatch cleavage with T4 endonuclease VII. Proc Natl Acad Sci U S A 92:87–91
31. Nelson NC, Hammond PW, Matsuda E, Goud AA, Becker MM (1996) Detection of all single-base mismatches in solution by chemiluminescence. Nucleic Acids Res 24:4998–5003
32. Dubertret B, Calame M, Libchaber AJ (2001) Single-mismatch detection using gold-quenched fluorescent oligonucleotides. Nat Biotechnol 19:365–370
33. Kostrikis LG, Tyagi S, Mhlanga MM, Ho DD, Kramer FR (1998) Spectral genotyping of human alleles. Science 279:1228–1229
34. Marras SAE, Russell Kramer F, Tyagi S (1999) Multiplex detection of single-nucleotide variations using molecular beacons. Genet Anal Biomol Eng 14:151–156
35. Merouch M, Chow CS (1999) Thermodynamics of RNA hairpins containing single internal mismatches. Nucleic Acids Res 27:1118–1125
36. Bonnet G, Tyagi S, Libchaber A, Kramer FR (1999) Thermodynamic basis of the enhanced specificity of structured DNA probes. Proc Natl Acad Sci U S A 96:6171–6176
37. Tsourkas A, Behlke MA, Bao G (2002) Structure–function relationships of shared-stem and conventional molecular beacons. Nucleic Acids Res 30:4208–4215
38. Cantor CR (1980) Biophysical chemistry: Part III: The behavior of biological macromolecules. WH Freeman & Co, San Francisco
39. Marras SAE, Kramer FR, Tyagi S (2002) Efficiencies of fluorescence resonance energy transfer and contact-mediated quenching in oligonucleotide probes. Nucleic Acids Res 30:e122
40. Demidov VV, Frank-Kamenetskii MD (2004) Two sides of the coin: affinity and specificity of nucleic acid interactions. Trends Biochem Sci 29:62–71
41. Lomakin A, Frank-Kamenetskii MD (1998) A theoretical analysis of specificity of nucleic acid interactions with oligonucleotides and peptide nucleic acids (PNAs). J Mol Biol 276:57–70
42. Kuhn H, Demidov VV, Coull JM, Fiandaca MJ, Gildea BD, Frank-Kamenetskii M (2002) Hybridization of DNA and PNA molecular beacons to single-stranded and double-stranded DNA targets. J Am Chem Soc 124:1097–1103
43. Tan L, Li Y, Drake TJ, Moroz L, Wang K, Li J, Munteanu A, Yang CJ, Martinez K, Tan W (2005) Molecular beacons for bioanalytical applications. Analyst 130:1002–1005
44. Goel G, Kumar A, Puniya A, Chen W, Singh K (2005) Molecular beacon: a multitask probe. J Appl Microbiol 99:435–442
45. Bonnet G, Libchaber A (1999) Optimal sensitivity in molecular recognition. Phys A 263:68–77

46. Wang K, Tang Z, Yang CJ, Kim Y, Fang X, Li W, Wu Y, Medley CD, Cao Z, Li J (2009) Molecular engineering of DNA: molecular beacons. Angew Chem Int Ed 48:856–870
47. Bratu DP, Cha BJ, Mhlanga MM, Kramer FR, Tyagi S (2003) Visualizing the distribution and transport of mRNAs in living cells. Proc Natl Acad Sci U S A 100:13308–13313
48. Beisel CL, Bayer TS, Hoff KG, Smolke CD (2008) Model-guided design of ligand-regulated RNAi for programmable control of gene expression. Mol Syst Biol 4:224
49. Kim JR, Ostermeier M (2006) Modulation of effector affinity by hinge region mutations also modulates switching activity in an engineered allosteric TEM1 [beta]-lactamase switch. Arch Biochem Biophys 446:44–51
50. Vallée-Bélisle A, Ricci F, Plaxco KW (2009) Thermodynamic basis for the optimization of binding-induced biomolecular switches and structure-switching biosensors. Proc Natl Acad Sci U S A 106:13802–13807
51. Tsourkas A, Behlke MA, Bao G (2002) Hybridization of 2′-O-methyl and 2′-deoxy molecular beacons to RNA and DNA targets. Nucleic Acids Res 30:5168–5174
52. Yao G, Fang X, Yokota H, Yanagida T, Tan W (2003) Monitoring molecular beacon DNA probe hybridization at the single-molecule level. Chem Eur J 9:5686–5692
53. Nakano S, Kirihata T, Fujii S, Sakai H, Kuwahara M, Sawai H, Sugimoto N (2007) Influence of cationic molecules on the hairpin to duplex equilibria of self-complementary DNA and RNA oligonucleotides. Nucleic Acids Res 35:486–494
54. Chen C, Wang W, Wang Z, Wei F, Zhao XS (2007) Influence of secondary structure on kinetics and reaction mechanism of DNA hybridization. Nucleic Acids Res 35:2875–2884
55. Porschke D, Eigen M (1971) Co-operative non-enzymatic base recognition III. Kinetics of the helix–coil transition of the oligoribouridylic · oligoriboadenylic acid system and of oligoriboadenylic acid alone at acidic pH. J Mol Biol 62:361–364
56. Pörschke D, Uhlenbeck O, Martin F (1973) Thermodynamics and kinetics of the helix-coil transition of oligomers containing GC base pairs. Biopolymers 12:1313–1335
57. McConaughy BL, Laird CD, McCarthy BJ (1969) Nucleic acid reassociation in formamide. Biochemistry (NY) 8:3289–3295
58. Kohne DE, Levison SA, Byers MJ (1977) Room temperature method for increasing the rate of DNA reassociation by many thousandfold: the phenol emulsion reassociation technique. Biochemistry (NY) 16:5329–5341
59. Yao G, Tan W (2004) Molecular-beacon-based array for sensitive DNA analysis. Anal Biochem 331:216–223
60. Tsourkas A, Bao G (eds) (2001) Detecting mRNA transcripts using FRET-enhanced molecular beacons. ASME BED
61. Vijayanathan V, Thomas T, Sigal LH, Thomas T (2002) Direct measurement of the association constant of HER2/neu antisense oligonucleotide to its target RNA sequence using a molecular beacon. Antisense Nucleic Acid Drug Dev 12:225–233
62. Mhlanga MM, Vargas DY, Fung CW, Kramer FR, Tyagi S (2005) tRNA-linked molecular beacons for imaging mRNAs in the cytoplasm of living cells. Nucleic Acids Res 33:1902–1912
63. Santangelo PJ, Bao G (2007) Dynamics of filamentous viral RNPs prior to egress. Nucleic Acids Res 35:3602–3611
64. Martinez K, Estevez M, Wu Y, Phillips JA, Medley CD, Tan W (2009) Locked nucleic acid based beacons for surface interaction studies and biosensor development. Anal Chem 81:3448–3454
65. Wu Y, Yang CJ, Moroz LL, Tan W (2008) Nucleic acid beacons for long-term real-time intracellular monitoring. Anal Chem 80:3025–3028
66. Vester B, Wengel J (2004) LNA (locked nucleic acid): high-affinity targeting of complementary RNA and DNA. Biochemistry (NY) 43:13233–13241
67. Kurreck J, Wyszko E, Gillen C, Erdmann VA (2002) Design of antisense oligonucleotides stabilized by locked nucleic acids. Nucleic Acids Res 30:1911–1918
68. Aartsma-Rus A, Kaman W, Bremmer-Bout M, Janson A, Den Dunnen J, van Ommen GJB, van Deutekom J (2004) Comparative analysis of antisense oligonucleotide analogs for targeted DMD exon 46 skipping in muscle cells. Gene Ther 11:1391–1398

69. Romani A (2007) Regulation of magnesium homeostasis and transport in mammalian cells. Arch Biochem Biophys 458:90–102
70. Santangelo PJ (2010) Molecular beacons and related probes for intracellular RNA imaging. Wiley Interdiscip Rev Nanomed Nanobiotechnol 2:11–19
71. Koshkin AA, Nielsen P, Meldgaard M, Rajwanshi VK, Singh SK, Wengel J (1998) LNA (locked nucleic acid): an RNA mimic forming exceedingly stable LNA: LNA duplexes. J Am Chem Soc 120:13252–13253
72. Koshkin AA, Singh SK, Nielsen P, Rajwanshi VK, Kumar R, Meldgaard M, Olsen CE, Wengel J (1998) LNA (locked nucleic acids): synthesis of the adenine, cytosine, guanine, 5-methylcytosine, thymine and uracil bicyclonucleoside monomers, oligomerisation, and unprecedented nucleic acid recognition. Tetrahedron 54:3607–3630
73. Wang L, Yang CJ, Medley CD, Benner SA, Tan W (2005) Locked nucleic acid molecular beacons. J Am Chem Soc 127:15664–15665
74. Yoo SM, Keum KC, Yoo SY, Choi JY, Chang KH, Yoo NC, Yoo WM, Kim JM, Lee D, Lee SY (2004) Development of DNA microarray for pathogen detection. Biotechnol Bioprocess Eng 9:93–99
75. Gracey AY, Cossins AR (2003) Application of microarray technology in environmental and comparative physiology. Annu Rev Physiol 65:231–259
76. Fan J, Yang X, Wang W, Wood WH, Becker KG, Gorospe M (2002) Global analysis of stress-regulated mRNA turnover by using cDNA arrays. Proc Natl Acad Sci U S A 99:10611–10616
77. Fang Y, Frutos AG, Lahiri J (2003) Ganglioside microarrays for toxin detection. Langmuir 19:1500–1505
78. Sekar M, Bloch W, St John PM (2005) Comparative study of sequence-dependent hybridization kinetics in solution and on microspheres. Nucleic Acids Res 33:366–375
79. Gao Y, Wolf LK, Georgiadis RM (2006) Secondary structure effects on DNA hybridization kinetics: a solution versus surface comparison. Nucleic Acids Res 34:3370–3377
80. Peterson AW, Heaton RJ, Georgiadis RM (2001) The effect of surface probe density on DNA hybridization. Nucleic Acids Res 29:5163–5168
81. Vainrub A, Montgomery Pettitt B (2003) Surface electrostatic effects in oligonucleotide microarrays: control and optimization of binding thermodynamics. Biopolymers 68:265–270
82. Halperin A, Buhot A, Zhulina E (2005) Brush effects on DNA chips: thermodynamics, kinetics, and design guidelines. Biophys J 89:796–811
83. Tawa K, Knoll W (2004) Mismatching base-pair dependence of the kinetics of DNA–DNA hybridization studied by surface plasmon fluorescence spectroscopy. Nucleic Acids Res 32:2372–2377
84. Stillman BA, Tonkinson JL (2001) Expression microarray hybridization kinetics depend on length of the immobilized DNA but are independent of immobilization substrate. Anal Biochem 295:149–157
85. Glazer MI, Fidanza JA, McGall GH, Trulson MO, Forman JE, Frank CW (2007) Kinetics of oligonucleotide hybridization to DNA probe arrays on high-capacity porous silica substrates. Biophys J 93:1661–1676
86. Liu X, Tan W (1999) A fiber-optic evanescent wave DNA biosensor based on novel molecular beacons. Anal Chem 71:5054–5059
87. Wang H, Li J, Liu H, Liu Q, Mei Q, Wang Y, Zhu J, He N, Lu Z (2002) Label-free hybridization detection of a single nucleotide mismatch by immobilization of molecular beacons on an agarose film. Nucleic Acids Res 30:e61
88. Du H, Strohsahl CM, Camera J, Miller BL, Krauss TD (2005) Sensitivity and specificity of metal surface-immobilized molecular beacon biosensors. J Am Chem Soc 127:7932–7940
89. Du H, Disney MD, Miller BL, Krauss TD (2003) Hybridization-based unquenching of DNA hairpins on Au surfaces: prototypical "molecular beacon" biosensors. J Am Chem Soc 125:4012–4013
90. Sauthier ML, Carroll RL, Gorman CB, Franzen S (2002) Nanoparticle layers assembled through DNA hybridization: characterization and optimization. Langmuir 18:1825–1830
91. Dave N, Liu J (2010) Fast molecular beacon hybridization fast molecular bea-con hybridization. J Phys Chem B 114:15694–15699

Chapter 3
Application of Molecular Beacons in Real-Time PCR

Chunming Wang and Chaoyong James Yang

Abstract Real-time PCR or quantitative PCR (QPCR) is a powerful technique that allows measurement of PCR product while the amplification reaction proceeds. It incorporates the fluorescent element into conventional PCR as the calculation standard to provide a quantitative result. In this sense, fluorescent chemistry is the key component in QPCR. Till now, two types of fluorescent chemistries have been adopted in the QPCR systems: one is nonspecific probe and the other is specific. As a brilliant invention by Kramer et al. in 1996, molecular beacon is naturally suited as the reporting element in real-time PCR and has been adapted for many molecular biology applications. In this chapter, we briefly introduce the working principle of QPCR and overview different fluorescent chemistries, and then we focus on the applications of molecular beacons-like gene expression study, single-nucleotide polymorphisms and mutation detection, and pathogenic detection.

3.1 Introduction

Soon after MBs were first introduced, they found important application in real-time PCR [1, 2], which requires prompt signal production with high specificity. In this chapter, some basic principles of real-time PCR are discussed, and the applications of MBs in single-nucleotide polymorphisms (SNPs) genotyping and pathogenic detection are summarized.

C. Wang (✉)
College of Chemistry and Chemical Engineering, Xiamen University, Xiamen 361005, People's Republic of China
e-mail: cmwang@xmu.edu.cn

C.J. Yang
Department of Chemical Biology, College of Chemistry and Chemical Engineering, Xiamen University, Xiamen, P.R. China
e-mail: cyyang@xmu.edu.cn

C.J. Yang and W. Tan (eds.), *Molecular Beacons*, DOI 10.1007/978-3-642-39109-5_3,

3.2 Basic Principles of Real-Time PCR

The emergence of polymerase chain reaction (PCR) has enabled the amplification of one or many copies of DNA isolated from cell, tissue, or blood samples, subsequently sparking a revolution in biology. Nonetheless, standard PCR is only designed for positive recognition of the amplicons. Post-PCR analysis should be performed in order to characterize both the size and sequence of the product. A DNA gel electrophoresis to measure the size and the quantity of the amplicon is simple and inexpensive. However, it is not easy to discriminate among different amplicons having similar size by gel. Also, gel is not accurate enough to assess the starting target by visualizing the final PCR product with nucleic acid intercalating dyes, although, to some extent, more initial target leads to more PCR product. This can be explained by studying the kinetics of PCR [3].

A conventional PCR usually consists of three phases: exponential, nonexponential, and plateau. At the beginning, all the reagents are sufficient to guarantee good PCR efficiency, and amplification occurs in an exponential manner, with the initial DNA doubling after every cycle. As the cycles progress and reagents are consumed, the reaction starts to slow down. At this point, the PCR product is no longer doubled in every cycle, and nonexponential amplification dominates. Finally, after another several rounds of amplification, the PCR reaction no longer generates templates as a result of the lack of critical components in the reaction. This is commonly known as the plateau phase, or end point, of the PCR reaction [4]. In this sense, the final yield of PCR product is not primarily dependent on the target sequence in the sample.

Real-time PCR, which combines amplification and quantification of a target DNA molecule in a single assay, is now a routine and robust technique in molecular biology. By detecting the products generated every cycle in a "real-time mode" at the early exponential phase, real-time PCR can differentiate the three stages and also save time since it is not necessary to wait until the reaction terminates. Real-time PCR allows the PCR product to accumulate during every cycle to allow measurement through different fluorescent chemistries.

3.3 Fluorescent Measurements for Real-Time PCR

There are generally two types of measurements to acquire the fluorescent signal from the PCR product. The first type relies on DNA binding dyes, such as SYBR Green I, which binds nonspecifically to double-stranded DNA (dsDNA) and emits an enhanced fluorescence [5]. The other type is a probe-based approach. These probes are sequence specific, and most of them use fluorescence resonance energy transfer (FRET) as the reporting mechanism [6] and use the 5′-exonuclease activity of the DNA polymerase [7] to detect PCR amplification in real time.

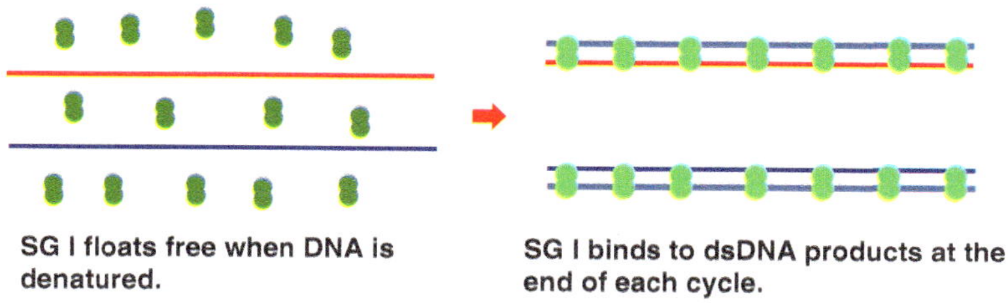

Fig. 3.1 A schematic of the working principle of SG I in real-time PCR

3.3.1 DNA Intercalating Dyes, SYBR Green I (SG I)

SG I shows a low fluorescence background when it is free in solution, while the fluorescence signal could increase up to 1,000-fold once it binds dsDNA (Fig. 3.1). This works universally for all dsDNA. Thus real-time PCR that utilizes SG I as reporter [5] is the most straightforward method by eliminating the complicated design of specific probes and reducing the cost of both time and money. As the PCR proceeds, more amplicons accumulate; accordingly, more SG I molecules are bound. The change in fluorescence can then be monitored using a thermocycler equipped with fluorescence detector. In other words, the fluorescence intensity is proportional to the amount of PCR products.

However, since SG I recognizes dsDNA in a nonspecific manner, nonspecific amplification products cannot be differentiated. Optimization of primers and template is therefore essential. First, the primers should be designed to generate amplicons with a suitable length (normally, 100–400 bp). Second, the concentration of primers should not be too high in order to achieve a high ratio of specific amplification versus primer-dimer signal. Third, the template should not include complicated secondary structures because this would contribute to the fluorescence signal. After real-time PCR is finished, a melting curve is usually recorded. If PCR generates a homogeneous sequence, only one transition point should be observed. Otherwise, nonspecific products or contaminants may exist. Most commonly, SG I is used for assays for which probe chemistry cannot be used or those assays which do not require high accuracy. SG I can also be used for optimization of primers prior to ordering the sequence-specific probe.

3.3.2 Probe-Based Chemistry

In probe-based chemistries, short oligonucleotides are used as an internal probe to hybridize with the region to be amplified. All these probes possess a quencher in close proximity to the reporter dye, where FRET occurs. In most cases, as the products form, the linkage between the quencher and dyes will be cut off, leading to a fluorescence enhancement. Compared to SG I, all probes are more specific since the hybridization only happens between probes and correct amplified products.

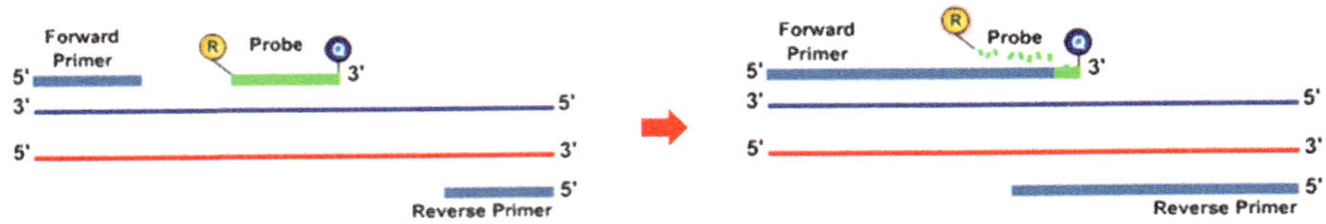

Fig. 3.2 Different states of TaqMan probe in real-time PCR: *left* – annealing; *right* – extension (Reprinted with permission from PREMIER Biosoft. Copyright ©1994–2013)

3.3.2.1 TaqMan Probe

TaqMan probe is a short single-stranded DNA (ssDNA) with a fluorophore at the 5′ end and a quencher at the 3′ end [8]. This DNA is complementary to the sequence within the template. Since the DNA is usually 20–30 bases in length, FRET is efficient when the probe is in free form. At this time, the fluorescence is quenched. It is only after the probe hybridizes to the template and is digested by Taq DNA polymerase (Taq polymerase is known to have 5′-exonuclease activity), as it extends the amplification primers, that the linkage between dye and quencher is cleaved, subsequently restoring fluorescence of the dye molecule (Fig. 3.2). Similar to SG I, the fluorescence increases in proportion to the amount of PCR products, since more probes will hybridize and will be cleaved. The advantages of the TaqMan probe include (1) analysis in real time without the need for post-PCR handling, thus reducing labor and cost; (2) specific hybridization-based detection, which eliminates nonspecific signal; and (3) labeling with different dyes with monitoring of different sequence amplifications in one tube. At the same time, however, TaqMan probes have some challenges, including (1) limited use in different assays and (2) limited design parameters. Normally, the probe should be designed close to the 5′ end to give a quick response. The length should be controlled in order to achieve sufficient FRET efficiency. In addition, guanine should not be placed next to the fluorescent dye since it is also an effective quencher.

3.3.2.2 FRET Hybridization Probes

The FRET probe system consists of two single-stranded fluorescent oligonucleotides such that probe 1 is labeled with a donor dye at the 3′ end, while probe 2 is labeled with an acceptor dye at the 5′ end [9, 10], typically having a 1–5 bases gap between them. During the annealing step, both probes will hybridize to the target, putting the donor in close proximity to the acceptor. The fluorescence signal of acceptor will then be detected, and the increase will be proportional to the products amplified (Fig. 3.3). Although this two-probe system gains specificity, it also increases the difficulty of hybridization.

Fig. 3.3 The working principle of FRET hybridization probes (Reprinted with permission from PREMIER Biosoft. Copyright ©1994–2013)

3.3.2.3 Molecular Beacons

The molecular beacons are the ideal hybridization-based probe for short oligonucleotide detection, and, thus, it is a suitable probe for real-time PCR [11]. In real-time PCR, MB hybridizes with template DNA at the annealing step and produces the fluorescent signal directly. Therefore, it does not need a polymerase with exonuclease activity, which is essential for the TaqMan probe. In the extension step, the polymerase will extend the sequence and displace the MB, returning it to the stem-loop conformation. In this case, the probe can be reused in the remaining cycles. MBs should be designed to hybridize 7–10 °C higher than primers, to ensure detection before primers are extended. Therefore, the stem should be just short enough to guarantee full hybridization, but not so short that can refold to the stem-loop structure after displacement of the molecular beacon by primer extension. Despite the difficulty in designing and optimizing a suitable MB, MB real-time PCR assays are simple, fast, sensitive, and accurate, allowing a high-throughput format and enabling the multiplexing detection in one tube using different labeling probes. MB-based PCR technique has been widely used in SNP analysis, real-time nucleic acid detection and quantitation, allele discrimination, and other clinical assays.

3.3.2.4 Scorpion Probe

The Scorpion probe is similar to molecular beacons in that it consists of a stem-loop structure when it is in free form. However, the Scorpion incorporates a primer into the sequence at the 3′ end, next to the quencher, via a blocker [12]. This blocker is a non-amplified monomer, which prevents the PCR from reading through the probe. In the extension stage, the polymerase binds the primer and synthesizes the complementary strand of target sequence just as it works in regular PCR. During the annealing step in next cycle, the loop will hybridize to the complementary strand within the same DNA. This separates the fluorophore and quencher, and an enhanced fluorescent signal is instantaneously observed (Fig. 3.4). Because the probe and primer are in the same molecule, the reaction kinetics is extremely fast. Also, intramolecular interaction is more favorable than intermolecular hybridization. This enables the Scorpion probe to provide a higher signal than other bimolecular systems, including either TaqMan or MB, but the design of the Scorpion is more difficult. Specifically, it reduces flexibility in probe

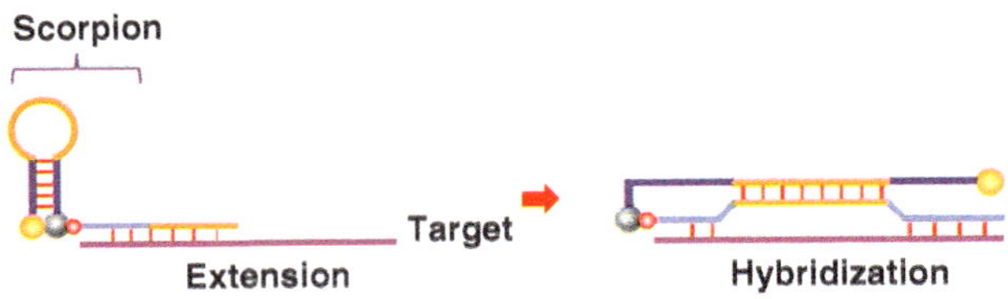

Fig. 3.4 The working principle of the Scorpion probe (Reprinted with permission from PREMIER Biosoft. Copyright ©1994–2013)

design where the loop should be engineered such that it is not too far from the complementary part to ensure high hybridization efficiency. Similarly, the stem should be long enough to stabilize the hairpin structure. The stem's Tm should be 5–10 °C higher than that of the primer-target hybrid.

3.4 MB Used in Real-Time PCR for SNPs and Mutation Detection Assays

The human genome consists of ten million single-nucleotide polymorphisms (SNPs). While most SNPs have no effect on health, some SNPs are believed to be related to the development of diseases [13]. Therefore, highly specific, simple, and accessible methods are needed for high-throughput SNPs detection (Fig. 3.5). MB-based assays provide a solution for screening SNPs in homogeneous assays [14]. Most of these assays require PCR to gain enough DNA targets, while MBs can specifically recognize these targets and present detectable signals in real time. Although other probes can also be used in real-time PCR, MBs have been demonstrated to be superior to them in certain aspects. For example, MBs have been proven to have better specificity than TaqMan probe in a detailed research report [15], and MBs have less complexity than Scorpion probe in design. Therefore, RT-PCR using MBs is perfect for SNPs analysis.

A key point for MBs in SNPs genotyping is to discriminate perfect match targets from single-base mismatch targets. The range of temperatures within which discrimination between the two targets is possible is wider for molecular beacons than it is for the corresponding linear probes. This is known as the window of discrimination, which is the basis for SNPs detection in homogeneous assays and is discussed at length in Chap. 4.

Early in 1998, 2 years after the MB was reported, Kramer et al. proposed the method of spectral genotyping human alleles using MB [11]. In their design, two MBs with different labeling were used: one specific for wild-type allele with green fluorophore and another for mutant allele labeled with red dye. The appearance of green, red, and both signals represented the homozygous wild type, homozygous mutant, and heterozygote, respectively.

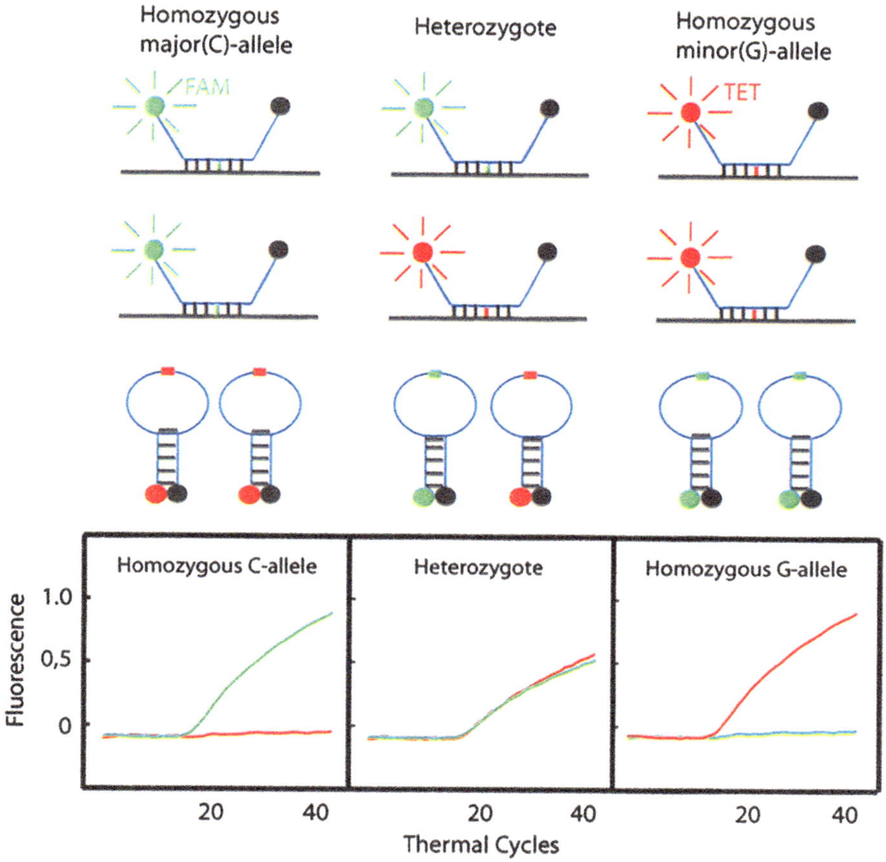

Fig. 3.5 Principle of spectral genotyping by PCR, exemplified by detection of a SNP in codon 325 of the estrogen receptor gene (Reprinted from Ref. [14]. Copyright 2001, with permission from Elsevier)

Later in 1998, in another report [16], they proved that MB-based sequence analysis could be adopted as an accurate assessment of DNA sequence. Five MBs, each complementary to a short fragment, were combined with 1–3 bases overlapped to span an 81 bp core region on the *rpoB* gene. This assay is simple and rapid. Most importantly, no contamination was observed since the tubes were not opened throughout the entire assay. Seventy-five clinical DNA isolates were correctly identified as drug susceptible or drug resistant. A broad range of point mutations, insertions, and deletions were detected successfully. Furthermore, in their paper in 1999 [17], up to four MBs, each with a different color, were used to explore four variants which differed from one another only by one base position. In four tubes, all the MBs were added with only one target variant. After PCR, only one fluorescence response was observed in each tube. This result indicates the extraordinary specificity of MB.

Since organic dyes often overlap each other in their emission spectra, 3–4 different dyes are the maximum that can be used at the same time. Another limitation comes from the instrument. Traditional thermal cyclers often have fixed excitation/emission filters. Under this condition, two dyes, which can be separate on a fluorometer, might not be able to be distinguished in PCR assays due to the lack of appropriate filters. Tyagi et al. proposed the construction of wavelength-shift MB [18], which emitted different fluorescent colors, but was excited with monochromatic light. This was realized by attaching a second fluorophore next to the fluorophore of a normal MB, which still contains a nonfluorescent quencher. In this design, one dye served as a harvester with strong absorption in the range of the light source. The other dye accepted the emission transferred from the harvester and emitted the desired color of fluorescence. This shift in emission spectrum is due to the fluorescence resonance energy transfer (FRET) from the harvester fluorophore to the emitter fluorophore. This only happened in opened probes that are bound to targets, and quencher has been separated from the fluorophores. By this method, the multiplex genetic analysis can be improved to be more simple and reliable.

3.5 MB Used in Real-Time PCR for Pathogenic Detection

Current techniques used to identify microbial pathogens usually rely on culturing and screening the samples to monitor the presence of pathogenic organisms, which are already well established. However, these suffer from a number of drawbacks. The assays are laborious, time-consuming, and expensive and require labile natural products [19]. More importantly, these routine tests do not directly characterize virulence factors. Efforts to overcome these problems in pathogenic detection have led to the development of DNA-based diagnosis. Today, culture-based methods for pathogen detection are rapidly being replaced by faster and more specific real-time PCR assays that discriminate different microorganisms based on a signal from specific nucleic acid sequences. Real-time PCR pathogen detection assays amplify target nucleic acid sequences from select microbes present in samples collected from complex biological environments. Specific amplification of target sequences is achieved by custom-designed primers and probes.

As pioneers in the study of molecular beacons, Kramer and colleagues described a multiplex MB assay to determine four pathogenic retroviruses [20]. Since then, scientists have applied this real-time PCR assay to detect all types of pathogenic organisms (Table 3.1).

Unlike assays that detect specific human DNA sequences present in samples, real-time PCR pathogenic detection assays must target genetic material from multiple microbial species in a single sample. This requires an assay which is capable of discriminating among the sequences from species of interest and other sequences from even the nearest evolutionary neighbors of the target species. Therefore, the detection of specific microorganisms requires the selection of an

Table 3.1 The utilization of MBs to detect different pathogens

Pathogens	Species	References
Bacteria and Fungus	*Aspergillus fumigatus*	[21]
	Bordetella pertussis	[22]
	Candida dubliniensis	[23]
	Chlamydophila felis	[24]
	Chlamydophila pneumoniae	[25]
	Clostridium	[26, 27]
	Escherichia coli	[28–31]
	Legionella pneumophila	[32]
	Mycobacterium tuberculosis	[33–44]
	Nitrifying bacteria	[45]
	Paracoccidioides brasiliensis	[46]
	Salmonella	[47–53]
	Scedosporiosis	[54]
	Staphylococcus	[55–58]
	Streptococcus	[59]
	Vibrio cholerae	[60]
Viruses	T-cell leukemia virus	[61]
	Adenovirus	[62, 63]
	Bluetongue virus	[64]
	Coronavirus	[65]
	Cytomegalovirus	[66]
	HAV	[67, 68]
	HBV	[69–74]
	HCV	[75–77]
	HPV	[78]
	Human immunodeficiency virus	[75, 79–83]
	Iridovirus	[84]
	Phage	[85]
	Plant virus	[86]
	Retrovirus	[20]
	Swine virus	[87]
	Syncytial virus	[88]
	West Nile virus	[89]

optimum target sequence to amplify. However, since many microbial sequences are unknown or have not yet been deciphered, the selection will be complicated and has no principle to rely on yet.

3.6 Conclusions

Real-time PCR represents one of the most important techniques in modern molecular biology, and it has become a routine and robust laboratory assay for gene expression analysis. In this chapter, different fluorescent chemistries applied in real-time PCR were summarized, reflecting the prominent role of MBs in this area. Based on the capacity to distinguish perfect match target from false targets, MBs are naturally suited for SNPs and mutation detection. Similarly, MBs can be used to detect all kinds of pathogens, which may only differ by a few bases in the gene sequences. With the increasing use of real-time PCR in gene transcription studies, disease-related diagnostics, and food safety assessment, MBs will continue to play an irreplaceable role in this field and aid the development of real-time PCR.

References

1. Heid CA, Stevens J, Livak KJ, Williams PM (1996) Real time quantitative PCR. Genome Res 6(10):986–994
2. Jung R, Soondrum K, Neumaier M (2000) Quantitative PCR. Clin Chem Lab Med 38(9):833–836
3. Higuchi R, Fockler C, Dollinger G, Watson R (1993) Kinetic PCR analysis: real-time monitoring of DNA amplification reactions. Biotechnology 11(9):1026–1030
4. Kainz P (2000) The PCR plateau phase – towards an understanding of its limitations. Biochim Biophys Acta 1494(1–2):23–27
5. Becker A, Reith A, Napiwotzki J, Kadenbach B (1996) A quantitative method of determining initial amounts of DNA by polymerase chain reaction cycle titration using digital imaging and a novel DNA stain. Anal Biochem 237(2):204–207
6. Hiyoshi M, Hosoi S (1994) Assay of DNA denaturation by polymerase chain reaction-driven fluorescent label incorporation and fluorescence resonance energy transfer. Anal Biochem 221(2):306–311
7. Holland PM, Abramson RD, Watson R, Gelfand DH (1991) Detection of specific polymerase chain reaction product by utilizing the 5′–3′ exonuclease activity of *Thermus aquaticus* DNA polymerase. Proc Natl Acad Sci U S A 88(16):7276–7280
8. Morris T, Robertson B, Gallagher M (1996) Rapid reverse transcription-PCR detection of hepatitis C virus RNA in serum by using the TaqMan fluorogenic detection system. J Clin Microbiol 34(12):2933–2936
9. Cardullo RA, Agrawal S, Flores C, Zamecnik PC, Wolf DE (1988) Detection of nucleic acid hybridization by nonradiative fluorescence resonance energy transfer. Proc Natl Acad Sci U S A 85(23):8790–8794
10. Lay MJ, Wittwer CT (1997) Real-time fluorescence genotyping of factor V Leiden during rapid-cycle PCR. Clin Chem 43(12):2262–2267
11. Kostrikis LG, Tyagi S, Mhlanga MM, Ho DD, Kramer FR (1998) Spectral genotyping of human alleles. Science 279(5354):1228–1229
12. Whitcombe D, Theaker J, Guy SP, Brown T, Little S (1999) Detection of PCR products using self-probing amplicons and fluorescence. Nat Biotechnol 17(8):804–807
13. What are single nucleotide polymorphisms (SNPs)? http://ghr.nlm.nih.gov/handbook/genomicresearch/snp
14. Mhlanga MM, Malmberg L (2001) Using molecular beacons to detect single-nucleotide polymorphisms with real-time PCR. Methods 25(4):463–471
15. Tapp I, Malmberg L, Rennel E, Wik M, Syvanen AC (2000) Homogeneous scoring of single-nucleotide polymorphisms: comparison of the 5′-nuclease TaqMan assay and Molecular Beacon probes. Biotechniques 28(4):732–738
16. Alland D, Piatek AS, Tyagi S, Pol AC, Telenti A, Miller LP, Kramer FR (1998) Molecular beacon sequence analysis for detecting drug resistance in *Mycobacterium tuberculosis*. Nat Biotechnol 16(4):359–363

17. Marras SAE, Kramer FR, Tyagi S (1999) Multiplex detection of single-nucleotide variations using molecular beacons. Genet Anal-Biomol E 14(5–6):151–156
18. Tyagi S, Marras SAE, Kramer FR (2000) Wavelength-shifting molecular beacons. Nat Biotechnol 18(11):1191–1196
19. Del Campo FJ, Lazcka O, Munoz FX (2007) Pathogen detection: a perspective of traditional methods and biosensors. Biosens Bioelectron 22(7):1205–1217
20. Vet JAM, Majithia AR, Marras SAE, Tyagi S, Dube S, Poiesz BJ, Kramer FR (1999) Multiplex detection of four pathogenic retroviruses using molecular beacons. Proc Natl Acad Sci U S A 96(11):6394–6399
21. Perlin DS, Nascimento AM, Goldman GH, Park S, Marras SAE, Delmas G, Oza U, Lolans K, Dudley MN, Mann PA (2003) Multiple resistance mechanisms among *Aspergillus fumigatus* mutants with high-level resistance to itraconazole. Antimicrob Agents Chemother 47(5):1719–1726
22. Poddar SK, Le CT (2001) *Bordetella pertussis* detection by spectrofluorometry using polymerase chain reaction (PCR) and a molecular beacon probe. Mol Cell Probe 15(3):161–167
23. Perlin DS, Park S, Wong M, Marras SAE, Cross EW, Kiehn TE, Chaturvedi V, Tyagi S (2000) Rapid identification of *Candida dubliniensis* using a species-specific molecular beacon. J Clin Microbiol 38(8):2829–2836
24. Helps C, Reeves N, Tasker S, Harbour D (2001) Use of real-time quantitative PCR to detect *Chlamydophila felis* infection. J Clin Microbiol 39(7):2675–2676
25. Gullsby K, Storm M, Bondeson K (2008) Simultaneous detection of *Chlamydophila pneumoniae* and *Mycoplasma pneumoniae* by use of molecular beacons in a duplex real-time PCR. J Clin Microbiol 46(2):727–731
26. Belanger SD, Boissinot M, Clairoux N, Picard FJ, Bergeron MG (2003) Rapid detection of *Clostridium difficile* in feces by real-time PCR. J Clin Microbiol 41(2):730–734
27. Paillard D, McKain N, Rincon MT, Shingfield KJ, Givens DI, Wallace RJ (2007) Quantification of ruminal *Clostridium proteoclasticum* by real-time PCR using a molecular beacon approach. J Appl Microbiol 103(4):1251–1261
28. Chen W, Fortin NY, Mulchandani A (2001) Use of real-time polymerase chain reaction and molecular beacons for the detection of *Escherichia coli* O157: H7. Anal Biochem 289(2):281–288
29. McKillip JL, Drake M (2000) Molecular beacon polymerase chain reaction detection of *Escherichia coli* O157:H7 in milk. J Food Prot 63(7):855–859
30. Ram S, Vajpayee P, Shanker R (2008) Rapid culture-independent quantitative detection of enterotoxigenic *Escherichia coli* in surface waters by real-time PCR with molecular beacon. Environ Sci Technol 42(12):4577–4582
31. Singh J, Batish VK, Grover S (2009) A molecular beacon-based duplex real-time polymerase chain reaction assay for simultaneous detection of *Escherichia coli* O157:H7 and *Listeria monocytogenes* in milk and milk products. Foodborne Pathog Dis 6(10):1195–1201
32. Templeton KE, Scheltinga SA, Sillekens P, Crielaard JW, van Dam AP, Goossens H, Claas EC (2003) Development and clinical evaluation of an internally controlled, single-tube multiplex real-time PCR assay for detection of *Legionella pneumophila* and other *Legionella* species. J Clin Microbiol 41(9):4016–4021
33. Chakravorty S, Aladegbami B, Motiwala AS, Dai Y, Safi H, Brimacombe M, Helb D, Alland D (2008) Rifampin resistance, Beijing-W clade-single nucleotide polymorphism cluster group 2 phylogeny, and the Rv2629 191-C allele in *Mycobacterium tuberculosis* strains. J Clin Microbiol 46(8):2555–2560
34. Dawes SS, Warner DF, Tsenova L, Timm J, McKinney JD, Kaplan G, Rubin H, Mizrahi V (2003) Ribonucleotide reduction in *Mycobacterium tuberculosis*: function and expression of genes encoding class Ib and class II ribonucleotide reductases. Infect Immun 71(11):6124–6131
35. Tyagi JS, Haldar S, Chakravorty S, Bhalla M, De Majumdar S (2007) Simplified detection of *Mycobacterium tuberculosis* in sputum using smear microscopy and PCR with molecular beacons. J Med Microbiol 56(10):1356–1362

36. Kumar P, Nath K, Rath B, Sen MK, Vishalakshi P, Chauhan DS, Katoch VM, Singh S, Tyagi S, Sreenivas V, Prasad HK (2009) Visual format for detection of *Mycobacterium tuberculosis* and *M. bovis* in clinical samples using molecular beacons. J Mol Diagn 11(5):430–438
37. Larsen MH, Vilcheze C, Kremer L, Besra GS, Parsons L, Salfinger M, Heifets L, Hazbon MH, Alland D, Sacchettini JC, Jacobs WR Jr (2002) Overexpression of inhA, but not kasA, confers resistance to isoniazid and ethionamide in *Mycobacterium smegmatis, M. bovis* BCG and *M. tuberculosis*. Mol Microbiol 46(2):453–466
38. Manganelli R, Dubnau E, Tyagi S, Kramer FR, Smith I (1999) Differential expression of 10 sigma factor genes in *Mycobacterium tuberculosis*. Mol Microbiol 31(2):715–724
39. Mathema B, Bifani PJ, Driscoll J, Steinlein L, Kurepina N, Moghazeh SL, Shashkina E, Marras SA, Campbell S, Mangura B, Shilkret K, Crawford JT, Frothingham R, Kreiswirth BN (2002) Identification and evolution of an IS6110 low-copy-number *Mycobacterium tuberculosis* cluster. J Infect Dis 185(5):641–649
40. Papaparaskevas J, Houhoula DP, Siatelis A, Tsakris A (2008) Molecular-beacon-based real-time PCR for detection and quantification of *Mycobacterium tuberculosis* DNA in clinical samples. J Clin Microbiol 46(9):3177–3178
41. Shi L, Jung YJ, Tyagi S, Gennaro ML, North RJ (2003) Expression of Th1-mediated immunity in mouse lungs induces a *Mycobacterium tuberculosis* transcription pattern characteristic of nonreplicating persistence. Proc Natl Acad Sci U S A 100(1):241–246
42. Singh A, Singh Y, Pine R, Shi L, Chandra R, Drlica K (2006) Protein kinase I of *Mycobacterium tuberculosis*: cellular localization and expression during infection of macrophage-like cells. Tuberculosis (Edinb) 86(1):28–33
43. Varma-Basil M, El-Hajj H, Colangeli R, Hazbon MH, Kumar S, Bose M, Bobadilla-del-Valle M, Garcia LG, Hernandez A, Kramer FR, Osornio JS, Ponce-de-Leon A, Alland D (2004) Rapid detection of rifampin resistance in *Mycobacterium tuberculosis* isolates from India and Mexico by a molecular beacon assay. J Clin Microbiol 42(12):5512–5516
44. Yesilkaya H, Meacci F, Niemann S, Hillemann D, Rusch-Gerdes S, Barer MR, Andrew PW, Oggioni MR (2006) Evaluation of molecular-Beacon, TaqMan, and fluorescence resonance energy transfer probes for detection of antibiotic resistance-conferring single nucleotide polymorphisms in mixed *Mycobacterium tuberculosis* DNA extracts. J Clin Microbiol 44(10):3826–3829
45. Harms G, Layton AC, Dionisi HM, Gregory IR, Garrett VM, Hawkins SA, Robinson KG, Sayler GS (2003) Real-time PCR quantification of nitrifying bacteria in a municipal wastewater treatment plant. Environ Sci Technol 37(2):343–351
46. Buitrago MJ, Merino P, Puente S, Gomez-Lopez A, Arribi A, Zancope-Oliveira RM, Gutierrez MC, Rodriguez-Tudela JL, Cuenca-Estrella M (2009) Utility of real-time PCR for the detection of *Paracoccidioides brasiliensis* DNA in the diagnosis of imported paracoccidioidomycosis. Med Mycol 47(8):879–882
47. Bhagwat AA, Patel J, Chua T, Chan A, Cruz SR, Aguilar GA (2008) Detection of *Salmonella* species in foodstuffs. Methods Mol Biol 429:33–43
48. Chen W, Martinez G, Mulchandani A (2000) Molecular beacons: a real-time polymerase chain reaction assay for detecting Salmonella. Anal Biochem 280(1):166–172
49. Hadjinicolaou AV, Demetriou VL, Emmanuel MA, Kakoyiannis CK, Kostrikis LG (2009) Molecular beacon-based real-time PCR detection of primary isolates of *Salmonella typhimurium* and *Salmonella enteritidis* in environmental and clinical samples. BMC Microbiol 9:97
50. Jyoti A, Ram S, Vajpayee P, Singh G, Dwivedi PD, Jain SK, Shanker R (2010) Contamination of surface and potable water in South Asia by Salmonellae: culture-independent quantification with molecular beacon real-time PCR. Sci Total Environ 408(6):1256–1263
51. Patel JR, Bhagwat AA, Sanglay GC, Solomon MB (2006) Rapid detection of Salmonella from hydrodynamic pressure-treated poultry using molecular beacon real-time PCR. Food Microbiol 23(1):39–46
52. Patel JR, Bhagwat AA (2008) Rapid real-time PCR assay for detecting Salmonella in raw and ready-to-eat meats. Acta Vet Hung 56(4):451–458

53. Uyttendaele M, Vanwildemeersch K, Debevere J (2003) Evaluation of real-time PCR vs automated ELISA and a conventional culture method using a semi-solid medium for detection of Salmonella. Lett Appl Microbiol 37(5):386–391
54. Castelli MV, Buitrago MJ, Bernal-Martinez L, Gomez-Lopez A, Rodriguez-Tudela JL, Cuenca-Estrella M (2008) Development and validation of a quantitative PCR assay for diagnosis of scedosporiosis. J Clin Microbiol 46(10):3412–3416
55. Chen L, Mediavilla JR, Oliveira DC, Willey BM, de Lencastre H, Kreiswirth BN (2009) Multiplex real-time PCR for rapid Staphylococcal cassette chromosome mec typing. J Clin Microbiol 47(11):3692–3706
56. Huletsky A, Giroux R, Rossbach V, Gagnon M, Vaillancourt M, Bernier M, Gagnon F, Truchon K, Bastien M, Picard FJ, van Belkum A, Ouellette M, Roy PH, Bergeron MG (2004) New real-time PCR assay for rapid detection of methicillin-resistant *Staphylococcus aureus* directly from specimens containing a mixture of staphylococci. J Clin Microbiol 42(5):1875–1884
57. Kreiswirth BN, Sinsimer D, Leekha S, Marras SAE, Koreen L, Willey B, Naidich S, Musser KA (2005) Use of a multiplex molecular beacon platform for rapid detection of methicillin and vancomycin resistance in *Staphylococcus aureus*. J Clin Microbiol 43(9):4585–4591
58. Warren DK, Liao RS, Merz LR, Eveland M, Dunne WM Jr (2004) Detection of methicillin-resistant *Staphylococcus aureus* directly from nasal swab specimens by a real-time PCR assay. J Clin Microbiol 42(12):5578–5581
59. Bergeron MG, Ke D (2001) New DNA-based PCR approaches for rapid real-time detection and prevention of group B streptococcal infections in newborns and pregnant women. Expert Rev Mol Med 3(27):1–14
60. Gubala AJ, Proll DF (2006) Molecular-beacon multiplex real-time PCR assay for detection of *Vibrio cholerae*. Appl Environ Microbiol 72(9):6424–6428
61. Besson G, Kazanji M (2009) One-step, multiplex, real-time PCR assay with molecular beacon probes for simultaneous detection, differentiation, and quantification of human T-cell leukemia virus types 1, 2, and 3. J Clin Microbiol 47(4):1129–1135
62. Claas EC, Schilham MW, de Brouwer CS, Hubacek P, Echavarria M, Lankester AC, van Tol MJ, Kroes AC (2005) Internally controlled real-time PCR monitoring of adenovirus DNA load in serum or plasma of transplant recipients. J Clin Microbiol 43(4):1738–1744
63. Poddar SK (1999) Detection of adenovirus using PCR and molecular beacon. J Virol Methods 82(1):19–26
64. Orru G, Ferrando ML, Meloni M, Liciardi M, Savini G, De Santis P (2006) Rapid detection and quantitation of Bluetongue virus (BTV) using a Molecular Beacon fluorescent probe assay. J Virol Methods 137(1):34–42
65. Hadjinicolaou AV, Farcas GA, Demetriou VL, Mazzulli T, Poutanen SM, Willey BM, Low DE, Butany J, Asa SL, Kain KC, Kostrikis LG (2011) Development of a molecular-beacon-based multi-allelic real-time RT-PCR assay for the detection of human coronavirus causing severe acute respiratory syndrome (SARS-CoV): a general methodology for detecting rapidly mutating viruses. Arch Virol 156(4):671–680
66. Jebbink J, Bai X, Rogers BB, Dawson DB, Scheuermann RH, Domiati-Saad R (2003) Development of real-time PCR assays for the quantitative detection of Epstein-Barr virus and cytomegalovirus, comparison of TaqMan probes, and molecular beacons. J Mol Diagn 5(1):15–20
67. Abd El Galil KH, El Sokkary MA, Kheira SM, Salazar AM, Yates MV, Chen W, Mulchandani A (2004) Combined immunomagnetic separation-molecular beacon-reverse transcription-PCR assay for detection of hepatitis A virus from environmental samples. Appl Environ Microbiol 70(7):4371–4374
68. Chen W, Yeh HY, Hwang YC, Yates MV, Mulchandani A (2008) Detection of hepatitis A virus by using a combined cell culture-molecular beacon assay. Appl Environ Microb 74(7):2239–2243
69. Lewin SR, Ribeiro RM, Walters T, Lau GK, Bowden S, Locarnini S, Perelson AS (2001) Analysis of hepatitis B viral load decline under potent therapy: complex decay profiles observed. Hepatology 34(5):1012–1020

70. Ntziora F, Paraskevis D, Haida C, Magiorkinis E, Manesis E, Papatheodoridis G, Manolakopoulos S, Beloukas A, Chryssoy S, Magiorkinis G, Sypsa V, Hatzakis A (2009) Quantitative detection of the M204V hepatitis B virus minor variants by amplification refractory mutation system real-time PCR combined with molecular beacon technology. J Clin Microbiol 47(8):2544–2550
71. Paraskevis D, Beloukas A, Haida C, Katsoulidou A, Moschidis Z, Hatzitheodorou H, Varaklioti A, Sypsa V, Hatzakis A (2010) Development of a new ultra sensitive real-time PCR assay (ultra sensitive RTQ-PCR) for the quantification of HBV-DNA. Virol J 7:57
72. Pas SD, Noppornpanth S, van der Eijk AA, de Man RA, Niesters HG (2005) Quantification of the newly detected lamivudine resistant YSDD variants of Hepatitis B virus using molecular beacons. J Clin Virol 32(2):166–172
73. Lai CL, Sum SSM, Wong DKH, Yuen MF, Yuan HJ, Yu J, Ho D, Zhang LQ (2004) Real-time PCR assay using molecular beacon for quantitation of hepatitis B virus DNA. J Clin Microbiol 42(8):3438–3440
74. Waltz TL, Marras S, Rochford G, Nolan J, Lee E, Melegari M, Pollack H (2005) Development of a molecular-beacon assay to detect the G1896A precore mutation in hepatitis B virus-infected individuals. J Clin Microbiol 43(1):254–258
75. Chung RT, Blackard JT, Komurian-Pradel F, Perret M, Sodoyer M, Smeaton L, St Clair JB, Chapman S, Taylor LE, Paranhos-Baccala G (2006) Intrahepatic cytokine expression is downregulated during HCV/HIV co-infection. J Med Virol 78(2):202–207
76. Komurian-Pradel F, Perret M, Deiman B, Sodoyer M, Lotteau V, Paranhos-Baccala G, Andre P (2004) Strand specific quantitative real-time PCR to study replication of hepatitis C virus genome. J Virol Methods 116(1):103–106
77. Morandi L, Ferrari D, Lombardo C, Pession A, Tallini G (2007) Monitoring HCV RNA viral load by locked nucleic acid molecular beacons real time PCR. J Virol Methods 140(1–2):148–154
78. Takacs T, Jeney C, Kovacs L, Mozes J, Benczik M, Sebe A (2008) Molecular beacon-based real-time PCR method for detection of 15 high-risk and 5 low-risk HPV types. J Virol Methods 149(1):153–162
79. Kostrikis LG, Touloumi G, Karanicolas R, Pantazis N, Anastassopoulou C, Karafoulidou A, Goedert JJ, Hatzakis A (2002) Quantitation of human immunodeficiency virus type 1 DNA forms with the second template switch in peripheral blood cells predicts disease progression independently of plasma RNA load. J Virol 76(20):10099–10108
80. Markowitz M, Lewin SR, Vesanen M, Kostrikis L, Hurley A, Duran M, Zhang L, Ho DD (1999) Use of real-time PCR and molecular beacons to detect virus replication in human immunodeficiency virus type 1 infected individuals on prolonged effective antiretroviral therapy. J Virol 73(7):6099–6103
81. Ho DD, Zhang LQ, Lewin SR, Markowitz M, Lin HH, Skulsky E, Karanicolas R, He YX, Jin X, Tuttleton S, Vesanen M, Spiegel H, Kost R, van Lunzen J, Stellbrink HJ, Wolinsky S, Borkowsky W, Palumbo P, Kostrikis LG (1999) Measuring recent thymic emigrants in blood of normal and HIV-1-infected individuals before and after effective therapy. J Exp Med 190(5):725–732
82. Gonzalez E, Bamshad M, Sato N, Mummidi S, Dhanda R, Catano G, Cabrera S, McBride M, Cao XH, Merrill G, O'Connell P, Bowden DW, Freedman BI, Anderson SA, Walter EA, Evans JS, Stephan KT, Clark RA, Tyagi S, Ahuja SS, Dolan MJ, Ahuja SK (1999) Race-specific HIV-1 disease-modifying effects associated with CCR5 haplotypes. Proc Natl Acad Sci U S A 96(21):12004–12009
83. Martinson JJ, Hong L, Karanicolas R, Moore JP, Kostrikis LG (2000) Global distribution of the CCR2-64I/CCR5-59653T HIV-1 disease-protective haplotype. AIDS 14(5):483–489
84. Wang XW, Ao JQ, Li QG, Chen XH (2006) Quantitative detection of a marine fish iridovirus isolated from large yellow croaker, *Pseudosciaena crocea*, using a molecular beacon. J Virol Methods 133(1):76–81

85. Kirs M, Smith DC (2007) Multiplex quantitative real-time reverse transcriptase PCR for F+−specific RNA coliphages: a method for use in microbial source tracking. Appl Environ Microbiol 73(3):808–814
86. Eun AJ, Wong SM (2000) Molecular beacons: a new approach to plant virus detection. Phytopathology 90(3):269–275
87. McKillen J, Hjertner B, Millar A, McNeilly F, Belak S, Adair B, Allan G (2007) Molecular beacon real-time PCR detection of swine viruses. J Virol Methods 140(1–2):155–165
88. O'Shea MK, Cane PA (2004) Development of a highly sensitive semi-quantitative real-time PCR and molecular beacon probe assay for the detection of respiratory syncytial virus. J Virol Methods 118(2):101–110
89. Lee DH, Mathew J, Pfahler W, Ma D, Valinsky J, Prince AM, Andrus L (2005) Individual donor nucleic acid amplification testing for detection of West Nile virus. J Clin Microbiol 43(10):5111–5116

Chapter 4
Molecular Beacons for Detection of Single-Nucleotide Polymorphisms

Mingxu You, Chaoyong James Yang, and Weihong Tan

Abstract Molecular beacons have been employed for the detection of single-nucleotide polymorphisms (SNPs) and allele discrimination because of their excellent selectivity in distinguishing single-nucleotide mutation. Moreover, since SNPs represent the most abundant class of the human genome alterations, they have also been used as biomarkers to identify complex disease-related genes and to perform pharmacogenomic analysis of drug response. In this chapter, current molecular beacon-based SNP genotyping methods, chemistries, and platforms are discussed. More specifically, the strategies of developing SNP detection approach based on molecular beacon-incorporated platforms, such as real-time polymerase chain reaction, endonucleases- or ribonuclease-dependent amplification methods, quantum dots or various quencher-based reporting systems, and nucleic acids' secondary-structure-opening beacon probes, are emphasized. Finally, current efforts in using molecular beacons for real human genome SNP analysis are summarized, providing a guideline for smarter use of molecular beacons for SNP genotyping.

M. You (✉)
Department of Chemistry, University of Florida, Gainesville, FL 32611, USA
e-mail: you@chem.ufl.edu

C.J. Yang
Department of Chemical Biology, College of Chemistry and Chemical Engineering, Xiamen University, Xiamen, P.R. China
e-mail: cyyang@xmu.edu.cn

W. Tan
Department of Biomedical Engineering and Department of Chemistry, Hunan University, Changsha, P.R. China

Department of Chemistry, University of Florida, Gainesville, FL, USA
e-mail: tan@chem.ufl.edu

C.J. Yang and W. Tan (eds.), *Molecular Beacons*, DOI 10.1007/978-3-642-39109-5_4,

4.1 Introduction

Single-nucleotide polymorphisms (SNPs) are single-nucleotide variations in the genome, accounting for 80–90 % of all human genetic variations. It has been demonstrated that more than ten million SNPs are present in the human genome, i.e., on average, one SNP occurs in every 500–1,000 base pairs [1]. As an important by-product of the Human Genome Project, SNPs are thought to be powerful genetic markers to assist in the identification of allele origin, the study of genetically associated diseases, and drug development [2, 3]. Driven by their biological and commercial importance, various methods have been developed by corporations and research groups to characterize and understand these SNP variations [4, 5].

Most current SNP genotyping methods rely on nucleic acid-based protocols [6], including allele-specific oligonucleotide (ASO) hybridization, primer extension, ligation [7], or endonuclease digestion [8]. ASO methods [9], which have been utilized for several decades, are based on the differences in probe-target hybridization stabilities between perfect-matched and mismatched targets. Initially, the development of SNP detection was hampered by the small amounts of nucleic acids present in clinical specimens [10]. This obstacle has been overcome with the help of enzyme-based amplification strategies, such as polymerase chain reaction (PCR) [11], which permits the direct amplification of target DNA sequence (in real time) to detect single-nucleotide mutations. To date, various detection platforms have been used for analyzing the products of each nucleic acid protocol, as noted above. Fluorescence measurements [12] or gel electrophoresis [13] are the most frequently used methods, while methods based on mass spectrometry [14], high-performance liquid chromatography (HPLC) [15], electrochemistry [16], or microgravimetry [17] have also been developed for specific conditions.

Although many SNP genotyping methods have been developed over the past two decades [10, 18], many challenges and opportunities for optimization are lying ahead, and further improvements are needed to lower cost and complexity, as well as to increase selectivity, speed, and sample throughput. With their high SNP distinguishing property, molecular beacon (MB)-based SNP probes have been developed with these goals in mind. In this chapter, we will discuss the current state of molecular beacon engineering methodologies for optimized SNP detection applications.

4.2 Target Specificity of Molecular Beacons

Any method for SNP identification must be capable of distinguishing point mutations with high selectivity. The mismatched base induces the loss of overall free energy of binding; however, as the number of nucleotides increases (as in the relatively long human genome), such free energy penalty represents a smaller and smaller fraction. Consequently, some traditional hybridization probes, such as linear DNA probes, are less efficient for SNP genotyping applications, and other highly selective probes are desirable.

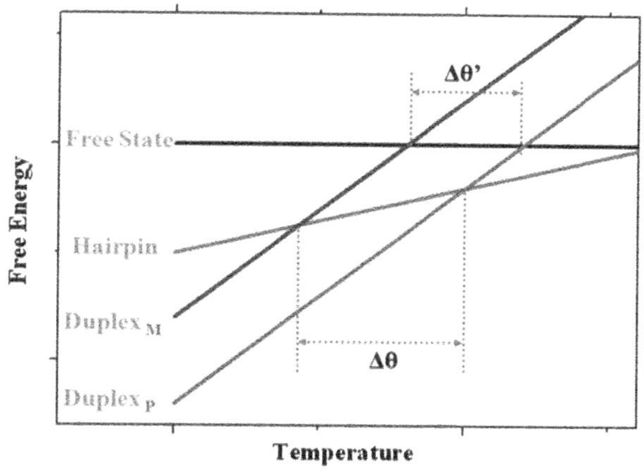

Fig. 4.1 Thermodynamic free energy calculation of MB-target-binding process, demonstrating that MB probes have a larger difference ($\Delta\theta$) between the melting temperature of perfectly matched duplexes (*green line*) and mismatched duplexes (*blue line*), when compared with linear probe ($\Delta\theta'$). The plot was modified from Ref. [21] (Color figure online)

For a probe depending on complementary bases, selectivity is determined by the phase-transition temperature differences between a perfect target-probe duplex and a mismatched helix. As the detailed theoretical thermodynamic studies illustrated in Chap. 2 (see also Fig. 4.1), MBs generally possess better target specificity compared with linear probes having a similar length, and MBs could distinguish mismatches over a wider temperature range [19–21]. Moreover, MBs with longer stems and shorter hairpin structures provide better target specificities to distinguish single-nucleotide mismatch. However, this leads to delayed recognition kinetics and reduced target-probe duplex stability [19]. In practice, to further improve target specificity, the mismatch site is usually designed so that it is positioned in the center of the loop domain.

4.3 Molecular Beacon-Based Real-Time PCR Detection of SNPs

As mentioned above, PCR amplification allows detection of trace amounts of SNPs in target DNA sequences. The combination of sensitive PCR methodology with selective MB probe design has generated a powerful method for real-time base-mutation analysis. Without the requirement of separation or purification steps, molecular beacons are very suitable for this application.

First, MBs can be incubated directly in the PCR reaction mixture to follow the reaction progress in real time. MBs have relatively rapid state-switching kinetics; after denaturing at a high temperature, they quickly return to hairpin structures at a lower temperature. The intramolecular interaction allows the renaturation of MBs to be finished in a much shorter time (<1 μs [20]) than that needed to sense the temperature effect. Thus, the progress of PCR can be monitored in real time by fluorescence readout (Fig. 4.2a). In one study [21], MBs were used for point mutation detection in a methylenetetrahydrofolate reductase (MTHFR) gene, which

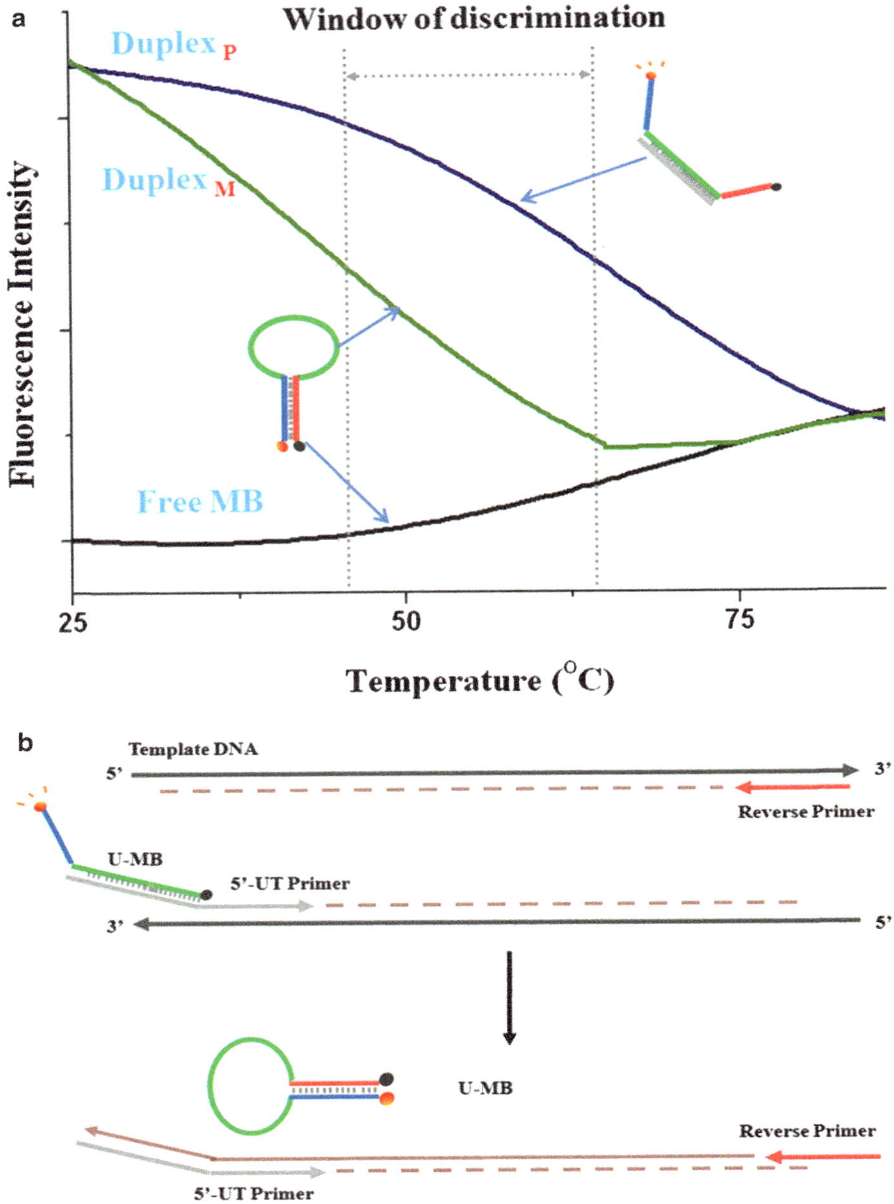

Fig. 4.2 MB-based real-time PCR assay: (**a**) fluorescence of MBs during PCR. SNPs analysis could be achieved within the temperature ranges termed "window of discrimination"; (**b**) schematic representation of 5′-UT primer and universal MB-based real-time PCR

represents an important marker for cardiovascular disease and neural tube defects. As the strands synthesized during PCR cycles accumulated, the number of MB probes that bound to the targets increased, leading to more intense fluorescence from the open fraction of MBs. The target specificity of the method was confirmed by the target-selective signaling from the design of molecular beacons. Thus, compared with other fluorescence-based hybridization methods (e.g., *TaqMan®* Gene Expression Assays [22]), MBs provide more reliable genotyping results. The entire procedure can be carried out in a closed-tube system, thereby avoiding contamination and achieving automation. MB-based real-time PCR has also proven useful for epidemiological studies [23]. Antifolate resistance-associated S108N point mutation could be successfully detected, and 100 African clinical isolates were tested using this new rapid and reliable method.

Second, real-time multiplex SNP analysis is possible by using different fluorophore-labeled molecular beacons in separate tubes or even in a single homogeneous system [24–26], as explained in detail in Chap. 3 and Fig. 4.2b.

In the practical utilization of MB-based PCR assay, the design of primers and beacons should also be rationally optimized. In general, the loop region of the MB is designed with a melting temperature slightly higher than the annealing temperature (T_{anneal}) of the PCR procedure (usually 18–25 bases) in order to ensure the stability of probe-target hybrids during signal detection process (i.e., annealing in the assay). The stem region is typically selected to dissociate at a temperature about 7–10 °C higher than the T_{anneal}. For example, if the $T_{\text{anneal}} = 55$ °C, then the melting temperature for the loop and stem region of MBs should be around 57 °C and 62–65 °C, respectively [11]. The design of PCR primers is also critical to obtain an optimum SNP-selective signal. Based on experience, 20–30 bases of a primer end should be used to flank the probe sequence. PCR products with a length of no more than 250 bp are generally used to design primers, since longer sequence may cause diminished signals.

As mentioned above, primer and beacon design can be important for PCR-based SNP analysis. In one study, Li et al. [27] designed a real-time PCR technique using a universal molecular beacon (U-MB) and a 5′-universal template primer (5′-UT primer). The 5′-UT primer was designed with a tail sequence (e.g., $(CT)_{10}$ CCGGG) to hybridize with the loop sequence (e.g., $(AG)_{10}$) and 5′-arm sequence (e.g., CCCGG) of the U-MB. As the number of PCR cycles increases, more new DNA fragments with 5′-UT primer tail will be synthesized. As the reverse primer extends, the hybridized U-MB will be replaced, resulting in a quenched fluorescence signal (Fig. 4.2b). This method has shown high sensitivity and point mutation discrimination ability, as demonstrated by detecting a single G mutation in codon 259 (AGA) of exon 7 of the p53 gene on breast cancer individuals [27]. The same U-MB could be used to identify different targets. It is worth noting that target specificity arises only from DNA polymerase activity, since the amplification of the mismatched template is usually significantly delayed relative to the perfectly matched one.

4.4 MBs Incorporated with Various Enzymatic Amplification Mechanisms

Other than the PCR technique, various types of enzyme-based amplification methods have been employed with molecular beacons for sensitive and specific SNP analysis. Among these enzymatic methods, DNA endonuclease-based approaches are most widely used. This type of isothermally amplified detection provides high sensitivity and a high potential for in vivo applications [28, 29]. However, the sequence-specific property of endonuclease limits the potential of enzymatic methods for general nucleic acid targeting. To overcome this difficulty, Kong et al. recently developed a MB-based "Y"-shaped probe with strong SNP identification ability [24]. As shown in Fig. 4.3, the Y-shaped junction consists of three parts: a signal MB probe, an assistant probe, and target DNA. The MB probe contains the nicking endonuclease site for a Nt. BbvCI enzyme, which recognizes dsDNA, but only hydrolyzes one specific strand (beacon). By changing the target DNA-bound sequence of the MB and assistant probes, while retaining the enzyme cleavage region, this smart design guarantees universal DNA target detection with the added potential of SNP selectivity. This efficient target-triggered amplification process can provide fast and sensitive responses.

Ribonuclease-based techniques are also attractive approaches for SNPs analysis. For example, using a MB with the loop region containing an SNP-specific ribonucleotide, Liu et al. reported a ribonuclease H (RNase H)-based technique for SNP detection [25]. If the MB probe is bound with a perfectly matched target, efficient cleavage by TthRNase HII will lead to fluorescence and gel signals. However, in the presence of a mismatched target, the cleavage efficiency was greatly reduced. Based on this strategy, the genotypes from ten individuals at 12 SNP sites were accurately determined, across a series of human leukocyte antigen (HLA) sites. Compared with other ribonucleases, such as RNase A, RNase 1, or RNase T, the use of RNase H allows more general applications, since other ribonucleases can only cleave certain types of mismatches [26].

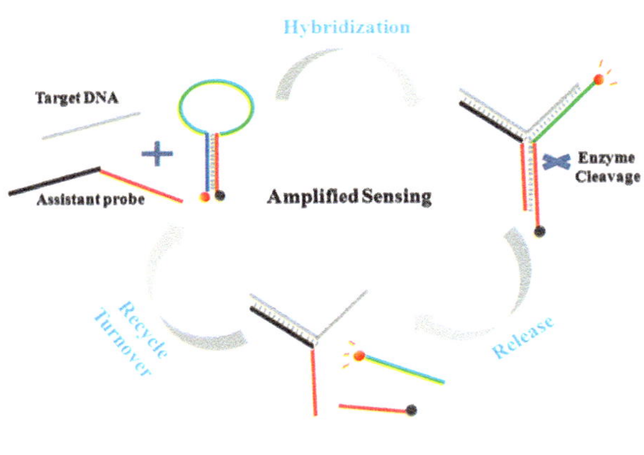

Fig. 4.3 Scheme of the Y-shaped junction; the nicking endonuclease only hydrolyzes the beacon probe, and SNP analysis was realized by the universal MB

An important benefit of nucleic acid chemistry is the ease of chemical modification; different functional groups or artificial nucleotides can be synthesized into oligonucleotide chains using a commercial DNA/RNA synthesizer. Peptide nucleic acid (PNA)-based beacon structures have been adapted for SNP genotyping. Komiyama et al. found that if a mismatch exists in the DNA sequence, the DNA could be efficiently hydrolyzed by nuclease S1 or Mung Bean nuclease after adding the complementary PNA. In contrast, fully matched DNA remained intact [30]. This behavior was attributed to conformational changes and physicochemical perturbations at the mismatch site. Moreover, the SNP sites could be visually detected using a 3,3′-diethylthiadicarbocyanine dye, which changes color in the presence of DNA/PNA duplexes. This type of PNA-nuclease system was later used in a PNA-MB system for SNP analysis [31]. Sensitive and selective SNP detection was demonstrated, and PNA beacons, which are resistant to naturally occurring enzymes, could be repeatedly used.

The enzymatic amplification method can be also hyphenated with electrochemical signals, instead of optical detection. In an interesting study, a biotin-labeled MB was immobilized on the gold electrode surface [32]. The MB-target hybridization event allowed the biotin group to interact with streptavidin-linked horseradish peroxidase (HRP), an enzyme whose activity was inhibited by stereo-hindrance in the absence of DNA target. Based on the electrochemical signal from the enzymatic product in the presence of substrate, sensitive DNA target detection and SNP analysis potential were demonstrated.

4.5 MBs with Modified Reporting Systems

As demonstrated in the above-mentioned HRP example, the reporting system of MBs has been modified in some efforts, as described in Chaps. 6 and 7. Interesting examples for SNP genotyping applications include the smart utilization of quantum dots (QDs) [33, 34], G-base quenching [35], gold nanoparticles [36, 37], and some nonfluorescence reporting systems [38].

Traditional organic fluophores have optical drawbacks, such as spectral crosstalk and the low resistance to photobleaching. In particular, spectral crosstalk limits their use for simultaneous multiple SNP detection. Quantum dots, which have narrow emission bands, broad absorptions, size-tunable photoluminescence spectra, and higher photostability, are attracting attention [39] for SNPs detection. Since the first use of quantum dot-based MBs (QD-MB) for single-base discrimination in 2004 [39], various multicolor QD-MB probes with different linkage strategies and quenchers have been reported with high sensitivity [40, 41]. In a recent example, the simultaneous detection of dual single-base mutations using QD-MB probes was achieved [40]. Using capillary electrophoresis for highly efficient separation of MB-target conjugates from unhybridized MBs further avoids the influence of false-positive signals. Sensitive target identification was achieved using 585- and 650-nm emitting CdTe QDs.

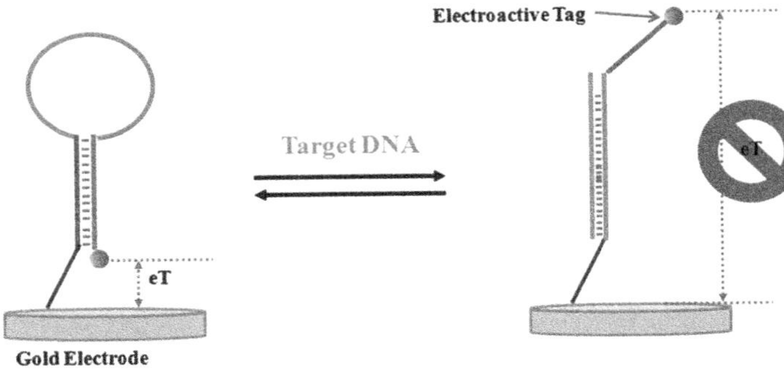

Fig. 4.4 Scheme of molecular beacon-based electrochemical probes for SNP genotyping

Improved quenching systems have also been employed, such as the gold nanoclusters. For example, 1.4 nm in diameter gold nanoparticles [35] have been used as a quencher for MBs. In comparison with organic quenchers (e.g., Dabcyl), gold clusters could improve the quenching efficiency as much as 100-fold. As a result, eightfold enhancement of single-nucleotide detection ability was achieved when compared to conventional MBs.

In another trial, gold nanoparticle-modified molecular beacons have also been used to develop a sensitive and low-cost method, which even allows visual SNPs analysis [35, 42]. In a recent study [43] with MB-functionalized gold nanoparticles, as low as 10 pM perfectly matched target DNA and mismatched DNA could be distinguished, based on the color change in a dry-reagent strip biosensor. This instrument-free, rapid (less than 30 min) and sensitive approach demonstrates the potential of conjugating MBs with nanomaterials to further optimize SNP genotyping methods.

Other than fluorescence or absorbance changes, electrochemical signals can be used to develop sensitive and inexpensive SNP genotyping systems. The most commonly used system is represented in Fig. 4.4. Molecular beacons are linked on one side with an electroactive tag, such as ferrocene or methylene blue, and with an electrode, such as gold, on the other side. The conformational change induced by the target hybridization event significantly alters the electron transfer between the electrode and the electroactive label, resulting in electrochemical signal alternations [42]. Because most of these electrochemical systems have been based on the "ON-to-OFF" mode, the decreased signal after target binding can influence the sensitivity of the method and may result in false-positives. In a recent study [44], Farjami et al. demonstrated a robust "OFF-to-ON" signal mode by truncating the MBs to 20 nts, resulting in strong selectivity for SNP analysis. As an example, this method could efficiently distinguish the SNP-containing sequences of TP53 DNA, a cancer biomarker gene, from normal genes.

4.6 Other Types of Molecular Beacon SNP Detection Systems

While the methods discussed above represent some of the most widely used approaches for MB-based detection of single-nucleotide mutations, the importance of SNP genotyping has also triggered some other smart designs in MB probes, and a few examples are described here.

Figure 4.5 shows a secondary-structure-folded nucleic acids (NA) probe using an Alzheimer's disease-related gene Tau-WT as the model target. Because Tau-WT has a stem structure, a single mutation can reduce stem ability and lead to an alternative mRNA splicing mechanism, eventually producing mutant proteins and forming nerve tissue-distracting filaments. Grimes et al. [45] demonstrated that a conventional MB probe, which contains a loop sequence complementary to Tau-WT, was unable to provide readable signals after binding with target. This result shows that MB probes may not be capable of detecting the presence of secondary-structure-folded targets. Instead, two target-specific adaptor DNA strands (f and m) were introduced and incubated with a universal piece of MB probe. The target-binding arm of strands f and m hybridize and unwind the secondary structure of the target. The resultant complex could further bind with MB to provide readout. A ninefold higher signal from perfectly matched target compared with mismatched DNA was demonstrated. The sequence of the MB is independent of the target sequence and can be synthesized in bulk amounts to reduce the cost. Similarly, some other multiple sequence systems have been developed by introducing "assistant strands." For example, PNA probes have been designed to act as "openers" to dehybridize double-stranded DNA targets, making them available for direct detection by MBs [46, 47].

In other examples, the stem or loop regions of MB probes have been engineered for advanced SNP detection. Lin et al. designed another universal MB based on a

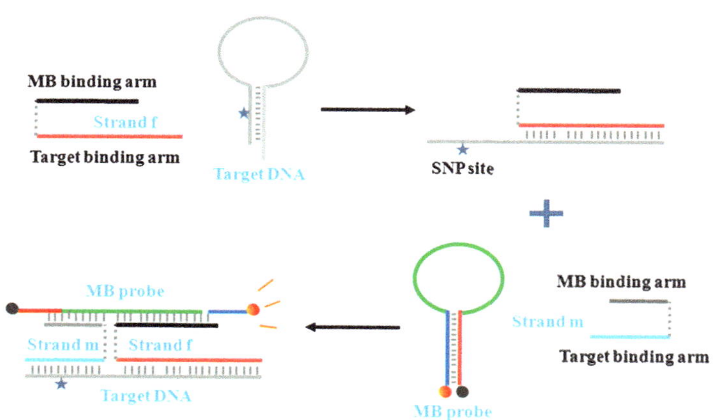

Fig. 4.5 Scheme of the secondary-structure-folded nucleic acids probe for SNP analysis

specific T-Hg^{2+}-T stem region [43]. The binding affinity of this MB probe could be easily adjusted by the concentration of Hg^{2+} ions. When compared with conventional MBs, a greater SNP discrimination ability was demonstrated by this design.

4.7 Conclusion and Future Perspectives

Since SNPs are extremely abundant in the human genome and are related to various diseases, a simple and accurate method to determine allele frequencies can provide important information, such as mapping genes that contribute to diseases and determining epidemiological patterns, as well as other genetic applications. The past few years have witnessed the rapid development of SNP genotyping technologies, and various SNP detection protocols have become available.

With their excellent mutation specificity, molecular beacons have been reliable probes for this type of biological application. This chapter has focused on the protocols currently incorporating MBs for SNP analysis, many leading to real human genome analysis (Table 4.1). However, no single protocol, whether based on

Table 4.1 Human gene SNP samples genotyped by molecular beacon assay

Genes name	SNP sites	Methods	Ref.
Exon 7 of *p53*	Codon *259*	Real-time PCR	[30]
MTHFR	*C677*T mutation		[27, 48]
DC-SIGN	54 SNPs		[10]
Estrogen receptor	Codon *325*		[11]
Chemokine receptor5	Codon *627*		[26]
DEFB1	Codon *44*		[49]
HFE	p. C*282*Y, p.H*63*D		[50]
GRK4;angiotensinogen; and aldosterone synthase	R*65*L, A*142*V and A*486*V; *6*G/A, M*235*T; *344*C/T		[51]
Estrogen receptor	Codon *10, 594, 325*		[52]
CCK, CYP17, DRD, HTR, MAOA, RS363	27 SNPs		[53]
Exon 4 of *IL-13*	A*2044*G	CE + PCR	[54]
Exon 6 of *XPD*	*A*-allele/*C*-allele	PNA + PCR	[55]
Human leukocyte antigen (HLA)	12 SNP sites	Ribonucleases	[34]
p53	Codon *273*	Ligase + ribonuclease	[56]
apoE3/E2 allele		PNA + nuclease	[37]
ApoE	Codon *112*	PNA + nuclease	[36]
Tau-WT	Tau/*0*C and Tau/*1*A	Secondary structure folding	[45]
rs1490413, rs876724, rs717302	rs14-G/A, rs71-G/A, rs87-C/T	Secondary structure folding	[57]
TP53	Codon *175*	Electrochemistry	[44]
p53	*HSC4, Ca9* mutant	DNA chips on glass	[58]

MBs or other modalities, has been able to meet all research demands. Factors, such as cost, time consumption, instrumentation, sensitivity, reproducibility, accuracy, multiplexing, and throughput level, still require improvements. In the future, smarter use of MBs for SNP genotyping will continue to be investigated.

References

1. NCBI dbSNP. http://www.ncbi.nlm.nih.gov/snp/index.html. Accessed 21 July 2011
2. Ding C (2007) 'Other' applications of single nucleotide polymorphisms. Trends Biotechnol 25:279–283
3. Kruglyak L (1999) Prospects for whole-genome linkage disequilibrium mapping of common disease genes. Nat Genet 22:139–144
4. Hinds DA, Stuve LL, Nilsen GB, Halperin E, Eskin E, Ballinger DG, Frazer KA, Cox DR (2005) Whole-genome patterns of common DNA variation in three human populations. Science 307:1072–1079
5. Miller RD, Phillips MS, Jo I, Donaldson MA, Studebaker JF, Addleman N, Alfisi SV, Ankener WM, Bhatti HA, Callahan CE, Carey BJ, Conley CL, Cyr JM, Derohannessian V, Donaldson RA, Elosua C, Ford SE, Forman AM, Gelfand CA, Grecco NM, Gutendorf SM, Hock CR, Hozza MJ, Hur S, In SM, Jackson DL, Jo SA, Jung SC, Kim S, Kimm K, Kloss EF, Koboldt DC, Kuebler JM, Kuo FS, Lathrop JA, Lee JK, Leis KL, Livingston SA, Lovins EG, Lundy ML, Maggan S, Minton M, Mockler MA, Morris DW, Nachtman EP, Oh B, Park C, Park CW, Pavelka N, Perkins AB, Restine SL, Sachidanandam R, Reinhart AJ, Scott KE, Shah GJ, Tate JM, Varde SA, Walters A, White JR, Yoo YK, Lee JE, Boyce-Jacino MT, Kwok PY, and Project SCAF (2005) High-density single-nucleotide polymorphism maps of the human genome. Genomics 86:117–126
6. Sobrino B, Brión M, Carracedo A (2005) SNPs in forensic genetics: a review on SNP typing methodologies. Forensic Sci Int 154:181–194
7. Toubanaki DK, Christopoulos TK, Ioannou PC, Flordellis CS (2009) Identification of single-nucleotide polymorphisms by the oligonucleotide ligation reaction: a DNA biosensor for simultaneous visual detection of both alleles. Anal Chem 81:218–224
8. Lyamichev V, Mast AL, Hall JG, Prudent JR, Kaiser MW, Takova T, Kwiatkowski RW, Sander TJ, de Arruda M, Arco DA, Neri BP, Brow MA (1999) Polymorphism identification and quantitative detection of genomic DNA by invasive cleavage of oligonucleotide probes. Nat Biotechnol 17:292–296
9. Wallace RB, Shaffer J, Murphy RF, Bonner J, Hirose T, Itakura K (1979) Hybridization of synthetic oligodeoxyribonucleotides to phi chi 174 DNA: the effect of single base pair mismatch. Nucleic Acids Res 6:3543–3557
10. Barreiro LB, Henriques R, Mhlanga MM (2009) High-throughput SNP genotyping: combining tag SNPs and molecular beacons. Methods Mol Biol 578:255–276
11. Mhlanga MM, Malmberg L (2001) Using molecular beacons to detect single-nucleotide polymorphisms with real-time PCR. Methods 25:463–471
12. Wang H, Li J, Wang Y, Jin J, Yang R, Wang K, Tan W (2010) Combination of DNA ligase reaction and gold nanoparticle-quenched fluorescent oligonucleotides: a simple and efficient approach for fluorescent assaying of single-nucleotide polymorphisms. Anal Chem 82:7684–7690
13. Gaunt TR, Hinks LJ, Rassoulian H, Day IN (2003) Manual 768 or 384 well microplate gel 'dry' electrophoresis for PCR checking and SNP genotyping. Nucleic Acids Res 31:e48
14. Mengel-Jorgensen J, Sanchez JJ, Borsting C, Kirpekar F, Morling N (2004) Multiplex Y chromosome SNP genotyping using MALDI-TOF mass spectrometry. Prog Forensic Genet 10(1261):15–17

15. Hecker KH, Taylor PD, Gjerde DT (1999) Mutation detection by denaturing DNA chromatography using fluorescently labeled polymerase chain reaction products. Anal Biochem 272:156–164
16. Huang Y, Zhang YL, Xu X, Jiang JH, Shen GL, Yu RQ (2009) Highly specific and sensitive electrochemical genotyping via gap ligation reaction and surface hybridization detection. J Am Chem Soc 131:2478–2480
17. Patolsky F, Lichtenstein A, Willner I (2001) Detection of single-base DNA mutations by enzyme-amplified electronic transduction. Nat Biotechnol 19:253–257
18. Kwok PY, Chen X (2003) Detection of single nucleotide polymorphisms. Curr Issues Mol Biol 5:43–60
19. Tsourkas A, Behlke MA, Rose SD, Bao G (2003) Hybridization kinetics and thermodynamics of molecular beacons. Nucleic Acids Res 31:1319–1330
20. Tyagi S, Kramer FR (1996) Molecular beacons: probes that fluoresce upon hybridization. Nat Biotechnol 14:303–308
21. Giesendorf BA, Vet JA, Tyagi S, Mensink EJ, Trijbels FJ, Blom HJ (1998) Molecular beacons: a new approach for semiautomated mutation analysis. Clin Chem 44:482–486
22. Livak KJ (1999) Allelic discrimination using fluorogenic probes and the 5′ nuclease assay. Genet Anal 14:143–149
23. Durand R, Eslahpazire J, Jafari S, Delabre JF, Marmorat-Khuong A, di Piazza JP, Le Bras J (2000) Use of molecular beacons to detect an antifolate resistance-associated mutation in Plasmodium falciparum. Antimicrob Agents Chemother 44:3461–3464
24. Kong RM, Zhang XB, Zhang LL, Huang Y, Lu DQ, Tan W, Shen GL, Yu RQ (2011) Molecular beacon-based junction probes for efficient detection of nucleic acids via a true target-triggered enzymatic recycling amplification. Anal Chem 83:14–17
25. Liu XP, Hou JL, Liu JH (2010) A novel single nucleotide polymorphism detection of a double-stranded DNA target by a ribonucleotide-carrying molecular beacon and thermostable RNase HII. Anal Biochem 398:83–92
26. Goldrick MM (2001) RNase cleavage-based methods for mutation/SNP detection, past and present. Hum Mutat 18:190–204
27. Li X, Huang Y, Guan Y, Zhao M, Li Y (2006) Universal molecular beacon-based tracer system for real-time polymerase chain reaction. Anal Chem 78:7886–7890
28. Kiesling T, Cox K, Davidson EA, Dretchen K, Grater G, Hibbard S, Lasken RS, Leshin J, Skowronski E, Danielsen M (2007) Sequence specific detection of DNA using nicking endonuclease signal amplification (NESA). Nucleic Acids Res 35:e117
29. Li JJ, Chu Y, Lee BY, Xie XS (2008) Enzymatic signal amplification of molecular beacons for sensitive DNA detection. Nucleic Acids Res 36:e36
30. Komiyama M, Ye S, Liang X, Yamamoto Y, Tomita T, Zhou JM, Aburatani H (2003) PNA for one-base differentiating protection of DNA from nuclease and its use for SNPs detection. J Am Chem Soc 125:3758–3762
31. Ye S, Miyajima Y, Ohnishi T, Yamamoto Y, Komiyama M (2007) Combination of peptide nucleic acid beacon and nuclease S1 for clear-cut genotyping of single nucleotide polymorphisms. Anal Biochem 363:300–302
32. Mao X, Jiang J, Xu X, Chu X, Luo Y, Shen G, Yu R (2008) Enzymatic amplification detection of DNA based on "molecular beacon" biosensors. Biosens Bioelectron 23:1555–1561
33. Kim JH, Morikis D, Ozkan M (2004) Adaptation of inorganic quantum dots for stable molecular beacons. Sens Actuators B 102:315–319
34. Li YQ, Guan LY, Wang JH, Zhang HL, Chen J, Lin S, Chen W, Zhao YD (2011) Simultaneous detection of dual single-base mutations by capillary electrophoresis using quantum dot-molecular beacon probe. Biosens Bioelectron 26:2317–2322
35. Saito Y, Mizuno E, Bag SS, Suzuka I, Saito I (2007) Design of a novel G-quenched molecular beacon: a simple and efficient strategy for DNA sequence analysis. Chem Commun (Camb):4492–4494

36. Dubertret B, Calame M, Libchaber AJ (2001) Single-mismatch detection using gold-quenched fluorescent oligonucleotides. Nat Biotechnol 19:365–370
37. Mao X, Xu H, Zeng Q, Zeng L, Liu G (2009) Molecular beacon-functionalized gold nanoparticles as probes in dry-reagent strip biosensor for DNA analysis. Chem Commun (Camb):3065–3067
38. Li D, Song S, Fan C (2010) Target-responsive structural switching for nucleic acid-based sensors. Acc Chem Res 43:631–641
39. Alivisatos AP (1996) Semiconductor clusters, nanocrystals, and quantum dots. Science 271:933–937
40. Kim JH, Chaudhary S, Ozkan M (2007) Multicolour hybrid nanoprobes of molecular beacon conjugated quantum dots: FRET and gel electrophoresis assisted target DNA detection. Nanotechnology 18:195105
41. Medintz IL, Berti L, Pons T, Grimes AF, English DS, Alessandrini A, Facci P, Mattoussi H (2007) A reactive peptidic linker for self-assembling hybrid quantum dot-DNA bioconjugates. Nano Lett 7:1741–1748
42. Nesterova IV, Erdem SS, Pakhomov S, Hammer RP, Soper SA (2009) Phthalocyanine dimerization-based molecular beacons using near-IR fluorescence. J Am Chem Soc 131:2432–2433
43. Lin YW, Ho HT, Huang CC, Chang HT (2008) Fluorescence detection of single nucleotide polymorphisms using a universal molecular beacon. Nucleic Acids Res 36:e123
44. Farjami E, Clima L, Gothelf K, Ferapontova EE (2011) "Off-on" electrochemical hairpin-DNA-based genosensor for cancer diagnostics. Anal Chem 83:1594–1602
45. Grimes J, Gerasimova YV, Kolpashchikov DM (2010) Real-time SNP analysis in secondary-structure-folded nucleic acids. Angew Chem Int Ed Engl 49:8950–8953
46. Kuhn H, Demidov VV, Coull JM, Fiandaca MJ, Gildea BD, Frank-Kamenetskii MD (2002) Hybridization of DNA and PNA molecular beacons to single-stranded and double-stranded DNA targets. J Am Chem Soc 124:1097–1103
47. Smolina IV, Demidov VV, Soldatenkov VA, Chasovskikh SG, Frank-Kamenetskii MD (2005) End invasion of peptide nucleic acids (PNAs) with mixed-base composition into linear DNA duplexes. Nucleic Acids Res 33:e146
48. Shi M, Caprau D, Dagle J, Christiansen L, Christensen K, Murray JC (2004) Application of kinetic polymerase chain reaction and molecular beacon assays to pooled analyses and high-throughput genotyping for candidate genes. Birth Defects Res A Clin Mol Teratol 70:65–74
49. Boniotto M, Hazbón MH, Jordan WJ, Lennon GP, Eskdale J, Alland D, Gallagher G (2004) Novel hairpin-shaped primer assay to study the association of the -44 single-nucleotide polymorphism of the DEFB1 gene with early-onset periodontal disease. Clin Diagn Lab Immunol 11:766–769
50. Alsmadi OA, Al-Kayal F, Al-Hamed M, Meyer BF (2006) Frequency of common HFE variants in the Saudi population: a high throughput molecular beacon-based study. BMC Med Genet 7:43
51. Bengra C, Mifflin TE, Khripin Y, Manunta P, Williams SM, Jose PA, Felder RA (2002) Genotyping of essential hypertension single-nucleotide polymorphisms by a homogeneous PCR method with universal energy transfer primers. Clin Chem 48:2131–2140
52. Täpp I, Malmberg L, Rennel E, Wik M, Syvänen AC (2000) Homogeneous scoring of single-nucleotide polymorphisms: comparison of the 5′-nuclease TaqMan assay and Molecular Beacon probes. Biotechniques 28:732–738
53. Myakishev MV, Khripin Y, Hu S, Hamer DH (2001) High-throughput SNP genotyping by allele-specific PCR with universal energy-transfer-labeled primers. Genome Res 11:163–169
54. Wang J, Wang W, Liu Y, Duo L, Huang L, Jiang X (2009) The method of single-nucleotide variations detection using capillary electrophoresis and molecular beacons. Mol Biol Rep 36:1903–1908

55. Petersen K, Vogel U, Rockenbauer E, Nielsen KV, Kølvraa S, Bolund L, Nexø B (2004) Short PNA molecular beacons for real-time PCR allelic discrimination of single nucleotide polymorphisms. Mol Cell Probes 18:117–122
56. Tang H, Yang X, Wang K, Tan W, Li H, He L, Liu B (2008) RNA-templated single-base mutation detection based on T4 DNA ligase and reverse molecular beacon. Talanta 75:1388–1393
57. Gerasimova YV, Hayson A, Ballantyne J, Kolpashchikov DM (2010) A single molecular beacon probe is sufficient for the analysis of multiple nucleic acid sequences. Chembiochem 11:1762–1768
58. Ohkubo A, Taguchi H, Seio K, Nagasawa H, Tsukahara T, Sekine M (2007) A new hydrophobic linker effective for the in situ synthesis of DNA-CPG conjugates as tools for SNP analysis. Tetrahedron Lett 48:5147–5150

Chapter 5
Molecular Beacons on Solid Surfaces

Tao Chen and Weihong Tan

Abstract In this chapter, immobilization of molecular beacons (MBs) on different bulk surfaces and nanomaterial surfaces is discussed. For use in large-scale parallel analysis with increased sensitivity at a pathogen-detector interface, MBs have been immobilized on different bulk solid support matrixes, including microscope glass slides, gold surfaces, and optical fibers. However, the performance of MBs on solid surfaces has been greatly compromised by severe surface effects. To solve this problem, various immobilization techniques and surface treatment approaches have been used to minimize interfacial issues. Since nanomaterials possess unique physical and chemical properties, they have also been utilized as solid supports to immobilize MBs. This chapter reviews recent progress in this area.

5.1 Introduction

For their use in large-scale parallel analysis with increased sensitivity at a pathogen-detector interface, it is convenient to first immobilize MBs onto a solid platform. Compared to MB assays performed in homogeneous phase, surface-immobilized MBs allow [1] simultaneous detection of multiple targets through spatial encoding. Moreover, the implementation of surface-immobilized MBs configured into an array of formats enables the seamless coupling to a microfluidic

T. Chen (✉)
Department of Chemistry, University of Florida, Gainesville, FL 32611, USA
e-mail: tchen51@ufl.edu

W. Tan
Department of Chemistry, University of Florida, Gainesville, FL, USA

Department of Biomedical Engineering and Department of Chemistry, Hunan University, Changsha 410082, P.R. China
e-mail: tan@chem.ufl.edu

C.J. Yang and W. Tan (eds.), *Molecular Beacons*, DOI 10.1007/978-3-642-39109-5_5,

system, which offers high levels of automation, as well as reduced sample and reagent consumption, leading to lower assay cost. Therefore, various solid surfaces have been studied as viable substrates for the attachment of MBs. In this chapter, we will mainly focus on such bulk solid surfaces as microscope glass slides, gold surfaces, and optical fibers. The immobilization of MBs on the surface of various nanomaterials, including gold nanoparticles (AuNPs), quantum dots (QDs), and magnetic nanoparticles (MNPs), will also be discussed.

5.2 MBs on Microscope Glass Slides

5.2.1 Non-covalent Immobilization through Biotin-(Strept)Avidin Interaction

5.2.1.1 Characteristics of Biotin-(Strept)Avidin Interaction

Avidin, a 67-kDa glycoprotein found in egg white, and streptavidin, a similar 60-kDa protein identified in *Streptomyces avidinii*, have extraordinarily high affinity to biotin, a 224.31-Da small molecule, also known as vitamin H [2, 3]. Their structures are shown in Fig. 5.1. The dissociation constant (K_d) of the biotin-avidin complex is on the order of 4×10^{-14} M [4], whereas the K_d of the biotin-streptavidin complex is measured to be around 10^{-15} M [5], making them among the strongest non-covalent interactions known in nature. In addition to this property, biotin-(strept)avidin interaction has some other unique characteristics that make it ideal as a general bridge system in immobilizing MBs onto various solid surfaces [6]:

(a) Once formed, the biotin-(strept)avidin complex is not disturbed by the presence of organic solvents, denaturants, detergents, chaotropes, proteolytic enzymes, changes in temperature, pH, or manipulations, such as multiple washings.
(b) Biotin-(strept)avidin interaction is highly specific, which ensures that the binding is only directed to the target of interest. However, in some applications, avidin has some nonspecific binding caused by its sugar moieties and high isoelectric point of around 10.

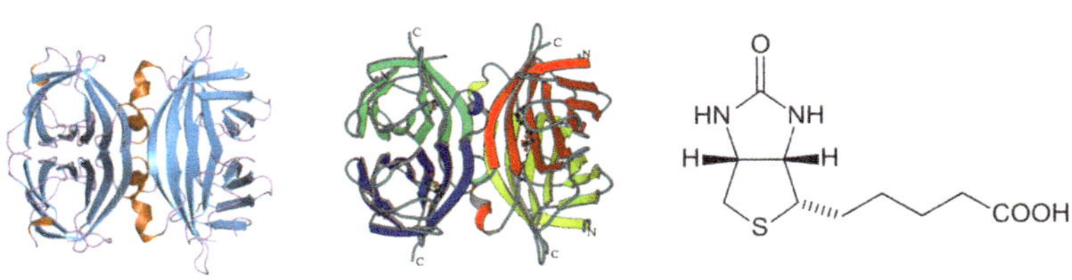

Fig. 5.1 Structures of avidin (*left*), streptavidin (*middle*) and biotin (*right*)

(c) (Strept)avidin possesses four binding sites per molecule, which makes its binding with multiple biotinylated molecules possible.
(d) When introduced into biologically active macromolecules, biotin does not usually affect their biological activity since it is just a small molecule. In addition, many biotinylation reagents are commercially available and extremely easy to use by following well-established protocols. For example, biotin moiety can be easily incorporated into different locations of an MB through phosphoramidite chemistry, in which DNA synthesis proceeds in the $3'$ to $5'$ direction, the opposite of biological systems.
(e) (Strept)avidin can be easily derivatized with various organic reagents for further conjugation with different compounds or solid supports.

5.2.1.2 Immobilization Methods

In order to immobilize MBs onto glass slides through biotin-(strept)avidin interaction, MBs should be biotinylated. Biotin moiety can be easily incorporated into MB at its $5'$-end using biotin phosphoramidite [7] or $3'$-end using biotin controlled pore glass (CPG) [8]. Biotin molecules can also be introduced in the middle of the MB sequence through the use of biotin-dT phosphoramidite [9].

Two approaches are commonly used for immobilizing MBs onto glass slides through biotin-(strept)avidin interaction: One is the direct method [10] and the other is the indirect method, as illustrated in Fig. 5.2 [9]. For direct immobilization of biotinylated MBs, highly concentrated (strept)avidin in appropriate buffer is first deposited onto glass slides. After multiple washings with buffer, the (strept)avidin on the surface is further stabilized by cross-linking with glutaraldehyde. Finally, biotinylated MBs are incubated with the activated surface after removal of unbound (strept)avidin. For indirect immobilization of biotinylated MBs, the glass slides are first incubated with highly concentrated BSA-biotin and then washed with buffer to remove unadsorbed BSA-biotin. The rest of the procedure is similar to the direct method, except no cross-linking is applied.

The direct method is simple and cost-effective. However, faster and more sensitive response of the immobilized MBs to their complementary targets is observed by using the indirect method [9]. In addition, while the immobilized MBs can only maintain their activity for 1 week if the direct method is used, immobilized MBs have up to a 3-week lifetime if the indirect method is used [9].

5.2.2 Covalent Immobilization through Siloxane-based Chemistry

Other than non-covalent immobilization of MBs onto glass slides through biotin-(strept)avidin interaction, covalent immobilization through siloxane-based

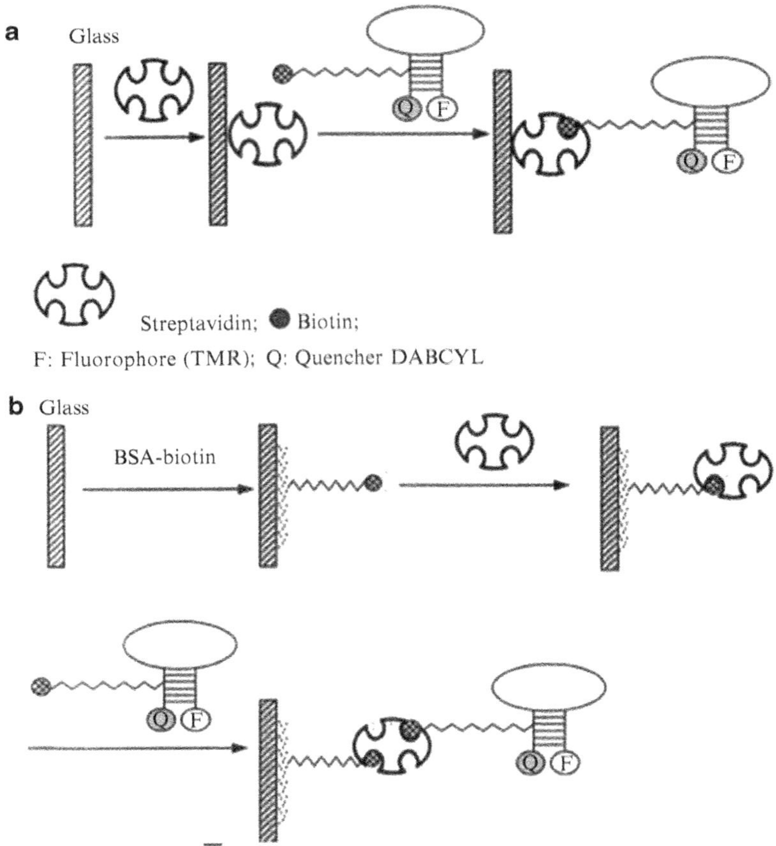

Fig. 5.2 Non-covalent immobilization of MBs onto glass slides through biotin-(strept)avidin interaction via (**a**) direct method and (**b**) indirect method (Reprinted with the permission from Ref. [9]. Copyright 2012 The Japan Society for Analytical Chemistry)

chemistry can also be employed [11–14]. As demonstrated in Fig. 5.3, MBs are commonly labeled with an amino group through phosphoramidite chemistry similar to that used in MB biotinylation. The glass slides are chemically treated with appropriate silanes to possess isothiocyanate [11, 12], aldehyde [13], or glycidoxy groups [14]. Taking the preparation of isothiocyanate group-bearing glass slides [12] as an example, cleaned glass slides are first immersed with 3-aminopropyltrimethoxysilane to introduce available amino groups. After multiple washings with buffer, they are furthered incubated with 1,4-phenylene diisothiocyanate solution in 10 % pyridine/dimethyl formamide to be labeled with isothiocyanate groups.

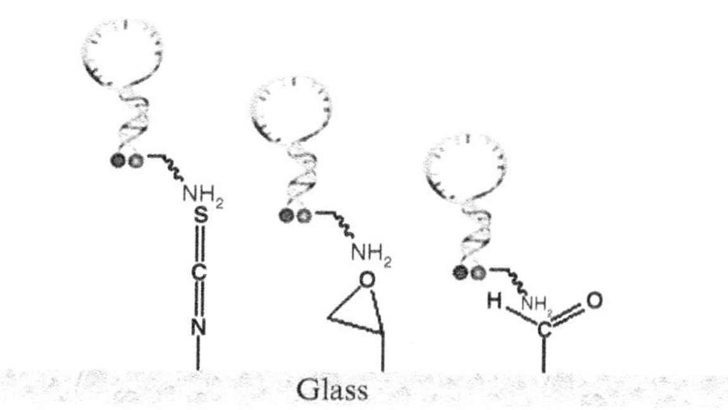

Fig. 5.3 Covalent immobilization of amino-labeled MBs onto modified glass slides through siloxane-based chemistry

5.2.3 Improving the Performance of MBs on Microscope Glass Slides

Although MBs have been non-covalently or covalently immobilized onto glass slides to develop DNA biochips, the immobilized MBs have much lower fluorescence enhancement than that in solution. Such reduced sensitivity of immobilized MBs has been attributed to their lower stability and reactivity on surfaces [15]. Based on their lower stability, immobilized MBs are partially open on the surface, and the lower reactivity then hinders effective opening after hybridization with their complementary targets. To address these issues, many new techniques for immobilizing MBs onto glass slides have been developed. For instance, both poly(thymidine/adenosine) (poly(T/A)) [16] and poly(ethylene glycol) (PEG) [17] spacing linkers are used to increase the distance between the MBs and surfaces, resulting in decreased surface effect and slightly improved MB performance.

Besides simply introducing PEG spacing linker, Tan and his colleagues have designed a special MB with alternating normal nucleic acid bases and locked nucleic acid (LNA) bases, aiming to enhance stability on the glass slide surface [7]. Compared to normal nucleic acid bases, LNA bases contain a bicyclic furanose unit locked in an RNA-mimicking sugar conformation [18]. The methylene bridge that connects the 2′-O and 4′-C of the ribose ring brings higher structural rigidity to the LNAs, thus preventing potential interactions between the immobilized MBs and the glass slide surface. Furthermore, LNAs have superior affinity and specificity to their complementary target when compared to that of DNA [19]. Using LNA-incorporated MBs on a glass slide surface was shown to be more stable, reproducible, selective, and robust when compared to normal MBs. Importantly, around 25-fold fluorescence enhancement was observed for the LNA-incorporated MBs on the glass slides. However, less than 10-fold enhancement was obtained for normal MBs.

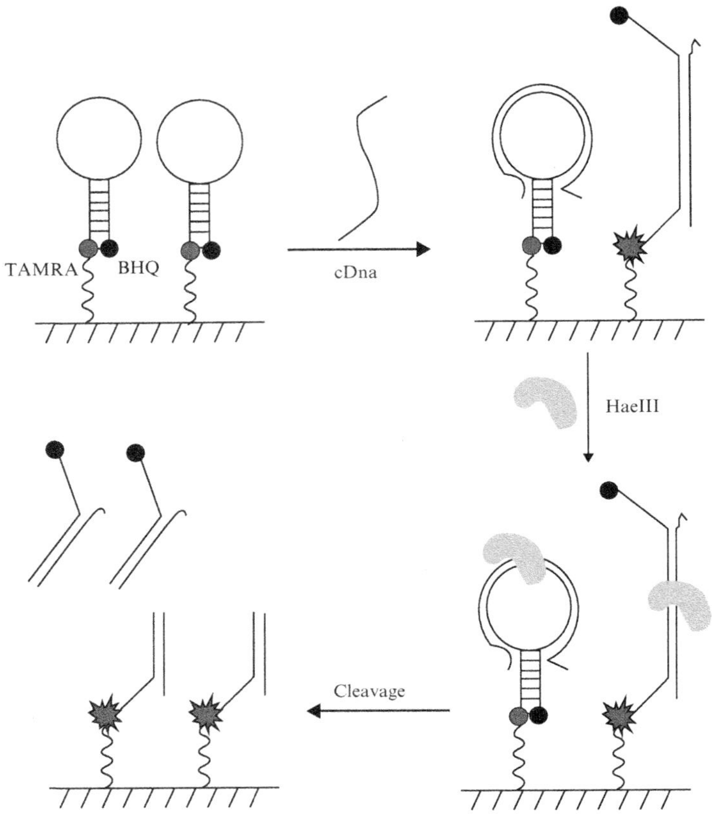

Fig. 5.4 Improving the performance of immobilized MBs through endonuclease cleavage. After hybridization with complementary target, only part of the surface-immobilized MBs open effectively. The endonuclease then cut all the hybridized duplex to release all fluorescence signals (Reprinted from Ref. [15]. Copyright 2012, with permission from Elsevier)

Considerable research effort has also been expended to increase the reactivity of immobilized MBs on surfaces. Wang et al. developed a novel MB bearing a restriction endonuclease recognition site at its loop structure and a poly(A) spacing linker between the MB and the surface, as shown in Fig. 5.4 [15]. The MBs were biotinylated at their 5′-end and immobilized onto the surface through indirect biotin-streptavidin interaction. After hybridization with their complementary targets, only some of the target-binding MBs were effectively opened, while the rest were still closed. Therefore, in the absence of endonuclease, only these fully opened MBs contributed to the final fluorescence signal. However, in the presence of endonuclease, the double-stranded target-binding MBs were specifically cleaved, whether they were open or closed, leading to much higher final fluorescence signal. Their results showed that improved fluorescence enhancement and sensitivity were accomplished by using this strategy.

Apart from incorporating different modification strategies to MBs, glass slides with different surface treatment approaches were also used to further improve the performance of surface-immobilized MBs. Among all these methods, coating with appropriate polymers has become the first choice. Different types of polymers have been used for this purpose, such as agarose [20], polyacrylamide [21], poly(methyl methacrylate) [1], and polypyrrole-polyvinylsulfonate [22]. Compared to plain glass slides, a polymer-coated surface provides a solution-like environment, rather than the typical heterogeneous liquid-solid interface, which can greatly reduce surface effect [20, 23]. Typically, immobilization of MBs onto these polymer-coated surfaces is realized by spotting MB solutions using a pin-based spotting robot. Lu et al. have successfully immobilized MBs onto agarose film-coated glass slides [20]. Their results showed that the immobilized MBs had high quenching efficiency, as well as excellent discrimination ratio for single-nucleotide mismatches. However, slow hybridization kinetics was observed for the agarose film-immobilized MBs due to the limited mass transport of the complementary targets through the polymer network.

5.3 MBs on Gold Surface

Even though microscope glass slides have been commonly used as a substrate for MB immobilization, they only play a passive role. Therefore, novel substrate materials with desired functions are in great demand. Among all potential substrates, the gold surface has greatly captured the interest of researchers, particularly since it can be used as a quenching agent through a resonance energy transfer or "contact quenching" process. To be immobilized onto gold substrate, quencher-free MBs can be used. Krauss et al. have developed two prototype MB biosensors by anchoring MBs labeled with only a fluorophore onto a gold surface through standard gold-thiol chemistry [24]. The quenching of the gold surface to both MBs was found to be more than 95 %, which is comparable to that obtained in solution-phase assays by using 4-((4′-(dimethylamino)phenyl)azo)benzoic acid (DABCYL) as the quencher. Based on the excellent quenching of the gold surface, more than 20-fold fluorescence enhancement was achieved, and complementary targets were detected down to 10 nM, using the gold-immobilized MBs. In addition, excellent selectivity of the immobilized MBs was noticed.

After demonstrating the feasibility of sensitively detecting complementary targets by using gold-immobilized, quencher-free MBs, Krauss et al. also applied this strategy for the detection of naturally occurring analytes, such as *Staphylococcus aureus femA* and *mecR* methicillin-resistant genes [25]. In order to ensure high hybridization efficiency between the surface-immobilized MBs and the complementary targets, some interstitial space between the probes is necessary [26]. In Krauss' work, the density of surface-immobilized MBs was optimized by changing the concentration of MBs in solution and the exposure time [25]. They discovered that simply using a low concentration of MB solution and a short incubation time did not necessarily result in low background fluorescence. Thus, they varied the coverage of surface-bound MBs by forming a two-component monolayer consisting of thiolated MBs and 3-mercapto-1-propanol (MP). Since MP was used as a spacer this time, low background fluorescence was observed. Sensitive detection of the desired targets by using the developed biosensors was achieved with more than 100-fold fluorescence enhancement. They also showed that the hybridization kinetics, sensitivity, and selectivity of the immobilized MBs were dependent on their specific sequence information. For example, MBs with longer hairpins possess more thermodynamically stable duplexes after hybridization to their complementary targets, resulting in higher sensitivity. At the same time, however, slower hybridization kinetics and less single-base mismatch discrimination are observed [25, 27].

In addition to its advantages, as outlined above, using a gold surface as a viable substrate has some drawbacks. Specifically, even though MBs can be easily attached to the gold surface via standard gold-thiol chemistry, nitrogen-based moieties in the DNA's backbone can also be chemically adsorbed onto the same surface, resulting in nonspecific adhesion of either MB themselves or their complementary targets. These nonspecific interactions can be minimized by employing blocking agents [1, 25]. Another problem involves nonuniform distribution of MBs on the gold surface, causing surface-induced aggregation that can affect the performance [28, 29].

5.4 MBs on Optical Fibers

An optical fiber is a flexible, transparent fiber made of pure glass that acts as a waveguide or "light pipe" to transmit light between the two ends of it. However, its properties are quite different from that of the microscope glass slide. An optical fiber generally has dimensions comparable to that of a human hair. Optical fibers are widely used in fiber-optic communications since light can travel along them over a long distance and at high bandwidths, with negligible loss and excellent immunity to electromagnetic interference. Specially designed optical fibers can also be used for a variety of other applications, including fiber lasers and sensors. Compared to other biosensors, optical fiber biosensors offer several advantages, small size, flexible geometry, remote working capability, and excellent immunity to noise, making them excellent candidates for environment monitoring and biomedical applications [30–34]. Optical fiber typically consists of a transparent core surrounded by a transparent cladding material with a lower index of refraction; therefore, light is kept in the core by total internal reflection, which forms the basis of fiber-optic evanescent wave biosensors [35].

A fiber-optic evanescent wave biosensor generally works in the following manner: First, fluorophores immobilized on or flowing around the core surface of an optical fiber are excited. Next, the fluorescence produced from the fluorophores is coupled back into the fiber core and finally transmitted to an optical detection system [35]. Many evanescent fiber sensors have been developed for sensitive detection of peptides, proteins, and DNA [36]. Tan et al. have developed a novel optical fiber evanescent wave DNA biosensor by immobilizing biotinylated MBs onto an optical fiber core surface through indirect biotin-streptavidin interaction [35]. The sensor is rapid, stable, reproducible, highly sensitive, and selective, and the concentration detection limit of the MB evanescent wave biosensor is 1.1 nM. They also investigated the influence of ionic strength relative to the hybridization kinetics of the immobilized MB to its complementary target. Their results showed that divalent cations, such as Mg^{2+}, play a more important role than monovalent cations, such as K^{+}, in speeding up the hybridization kinetics. Following this work, they then prepared an MB array and successfully utilized it for simultaneous analysis of multiple targets in the same solution [37].

5.5 MBs on Nanomaterial Surfaces

Nanomaterials are structures with a size of 100 nm or smaller in at least one dimension. Based on quantum effects that result from the large surface area to volume ratio, nanomaterials possess unique optical, electronic, magnetic, and mechanical properties [38]. Therefore, nanomaterials have generated widespread interest in biosensor development [39] and biomedical applications [40]. Hybrid materials composed of nanomaterials and biomolecules have recently been assembled and used in different scientific fields [41–43]. It is expected that the highly specific recognition abilities of biomolecules coupled with various properties of nanomaterials will make these composite materials more attractive in their applications. Therefore, MBs have been immobilized onto different nanomaterial surfaces to develop multifunctional probes. Here, we will focus on gold nanoparticles (AuNPs), quantum dots (QDs), and magnetic nanoparticles (MNPs).

5.5.1 MBs on AuNPs

Similar to bulk gold surfaces, AuNPs can also serve as quenching agents for immobilized fluorophores. Instead of using an organic quencher, such as DABCYL, a 1.4-nm AuNP was used as an inorganic quencher [44]. As demonstrated in Libchaber's work, a 25-base hairpin-structured synthetic oligonucleotide modified with primary amine at its 3′-end and a disulfide at its 5′-end was first synthesized. Then, an amine-reactive dye was coupled to the primary amine at the 3′-end, and the 1.4 nm AuNP with one *N*-propylmaleimide was attached to the activated disulfide at its 5′-end. Gel electrophoresis was used to analyze the reaction product, confirming the absence of nonspecific interactions between the DNA and the AuNP. In the absence of complementary target, the MB had negligible background. However, in the presence of target, significant fluorescence enhancement ensued. Compared to conventional MBs, their MB had sensitivity enhancement up to 100-fold and improvement in discrimination to single-base mismatched target up to 8-fold.

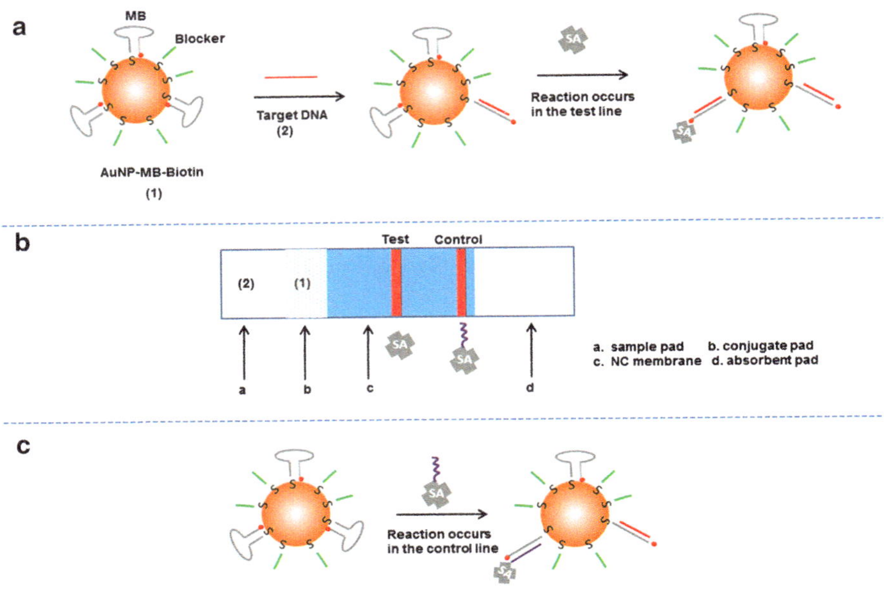

Fig. 5.5 Schematic illustration of the dry-reagent strip-type nucleic acid biosensor (*DSNAB*) based on an MB-modified AuNP probe and a lateral flow test strip. (**a**) In the test zone, sample solution containing complementary target rehydrates the MB-immobilized AuNPs, activating the biotin groups. Thus, the formed hybrids are captured by streptavidin to show a characteristic *red band*. (**b**) Diagram of the DSNAB. (**c**) In the control zone, excess MB-immobilized AuNPs are captured by the DNA probe (Reproduced from Ref. [47] by permission of The Royal Society of Chemistry)

The superior performance of the MB with an AuNP quencher comes from the better quenching ability of the AuNP compared to an organic quencher, especially in low-salt buffers.

In addition to their use as quenching agents, AuNPs have strong localized surface plasmon resonance [45], resulting in an intense red color (for particles less than 100 nm) [46]. Based on this property, Liu et al. have constructed a dry-reagent strip-type nucleic acid biosensor (DSNAB) based on an MB-modified AuNP probe and a lateral flow test strip [47]. As illustrated in Fig. 5.5, an MB modified with biotin at its 3′-end and thiol at its 5′-end was immobilized onto AuNPs in the presence of 11-mercaptoundecanol blockers, leading to the inactive state of biotin. Upon complementary target addition, duplex formation pulls the biotin away from the AuNP surface, and the activated biotin is able to react with streptavidin. A conventional immunochromatographic strip consists of four components: a sample pad, a conjugation pad, a nitrocellulose membrane, and an absorption pad. The MB-immobilized AuNPs were dispensed on the conjugation pad. Streptavidin and streptavidin-biotin-DNA (complementary to the MB) were dispensed on the nitrocellulose membrane to form the test and control zone. The sample solution containing complementary target DNA was applied on the sample pad. When the solution migrated along the strip, it first rehydrated the MB-immobilized AuNP to activate the biotin groups. The formed hybrids continued to migrate along the strip and were captured in the test zone by the pre-immobilized streptavidin, forming a characteristic red band to indicate the presence of the target. Then the excess MB-immobilized AuNPs were captured by the DNA probe, leading to a second characteristic red band to indicate that the DSNAB was working properly. Using this sensor, target DNA as low as 50 pM was visualized within 15 min with great discrimination to single-base mismatched target.

It is very challenging to deliver DNA across the plasma membrane barrier [48]; however, DNA-modified AuNPs can enter cells with high efficiency without external assistance [49]. Based on this concept, Tang et al. have developed a tumor mRNA-dependent MB-AuNP carrier for controlled release of doxorubicin and intracellular imaging. Doxorubicin is a fluorescent anticancer drug. It can intercalate into the double-stranded 5′-GC-3′ or 5′-CG-3′ sequences of DNA or RNA, resulting in quenched fluorescence and decreased cytotoxicity of doxorubicin [50]. Therefore, after immobilizing MBs onto AuNPs, doxorubicin was intercalated into the GC-containing stem part of the MB, constructing a reservoir for doxorubicin. The loop of the MB was complementary to cyclin D1 mRNA, which is highly expressed in breast cancer cell line SK-BR-3. Their results showed that selective killing was observed for target cells. In addition, the released doxorubicin in target cells illuminated the cell, enabling simultaneous imaging.

5.5.2 MBs on QDs

Organic dyes have been widely used for MB construction. However, organic dyes are sensitive to the physiological environment and vulnerable to photobleaching under normal imaging conditions [51]. In addition, two inherent properties prevent their use for multiplexed detection: (1) Organic dyes have relatively broad emission spectra, resulting in signal overlap from different dyes, and (2) only one organic dye can be optimally excited by a certain laser, thus requiring multiple lasers for multiplexed detection [52]. On the other hand, inorganic QDs are usually bright and stable, even under a relatively harsh environment [53, 54]. Importantly, QDs always emit the same light, no matter what excitation wavelength is used [52]. This means that only one excitation source is needed to see all the different colors from QDs. Therefore, other than organic dye, QD is also an exceptional candidate for MB construction, especially when simultaneous detection of multiple targets is needed.

To demonstrate the feasibility of using quantum dots for MB construction, Ozkan et al. have developed an MB with a ZnS-capped CdSe QDs as a fluorophore and DABCYL as a quencher [52]. The as-synthesized QDs were first functionalized with mercaptoacetic acid (MMA) to make them water soluble and possess carboxyl groups. Then amino group-modified MBs were attached onto their surface with the help of 1-ethyl-3-(3-dimethylaminopropyl)carbodiimide hydrochloride (EDC). Compared to an organic dye (FAM) MB, QD MBs had improved imaging lifetime.

However, since QDs are larger than DABCYL, the center-to-center spacing between QD and DABCYL was found to be larger when compared to the FAM/DABCYL case, leading to less quenching efficiency and thus less sensitivity of the QD MB. Only around five- to sixfold fluorescence enhancement was observed for the QD MB.

In order to further improve the performance of QD MBs, condition optimization of the system was carried out by many researchers. Batt et al. systematically studied the influence of linkage and quenching strategies to quantum dot MBs [55]. Two different linkage strategies were used to immobilize MBs onto QD surface: covalently link amino group-modified MBs onto carboxyl group bearing QDs and non-covalently link biotinylated MBs onto streptavidin-coated QDs. Since the hydrodynamic diameter of the streptavidin-coated QDs at 13.5 nm exceeded that of the carboxyl-bearing QDs at 7 nm, the amide-linked MBs showed a 57 % greater fluorescence enhancement over that of the streptavidin-linked MBs. Three different quenchers, Iowa Black, 1.4 nm AuNP, and DABCYL, were used to construct QD MBs. According to their results, Iowa Black- and 1.4-nm AuNP-quenched MBs exhibited approximately twofold greater fluorescence enhancement than that of DABCYL-quenched MBs. Putting all their results together, it can be concluded that it is very important to choose the proper linkage method and quencher moieties to develop high-quality QD MBs.

5.5.3 MBs on MNPs

The separation of single-base mismatched DNA/RNA strands in complex mixtures, such as cell lysates and tissue samples, is of great interest in disease diagnostics and biomedical studies. However, conventional DNA separation methods, such as high-pressure liquid chromatography and gel electrophoresis, are based on the hydrophobicity or the length of DNA, but lack sequence recognition capability. Recently, MBs have been developed for real-time DNA/RNA detection with high sensitivity and selectivity. More importantly, they have excellent capability to discriminate single-base mismatched DNA from complementary target DNA. Therefore, MBs can be used to solve the abovementioned problem if a suitable carrier between liquid and solid phases is provided.

Aside from having high surface-to-volume ratio and ultrasmall size, MNPs also obey Coulomb's law and can be manipulated by an external magnetic field gradient. Therefore, Tan et al. have used MNPs as carriers to immobilize MBs [56]. The silica-coated MNPs were first synthesized using a reverse microemulsion method [57]. Then they were further coated with a layer of avidin through electrostatic interaction. After cross-linking the avidin by glutaraldehyde, biotinylated MBs were finally immobilized onto the MNPs. The working principle of the system is illustrated in Fig. 5.6. First, single-base mismatched DNA and complementary target DNA were both captured. The melting temperature of the duplex between single-base mismatched DNA and MB is lower than complementary target DNA.

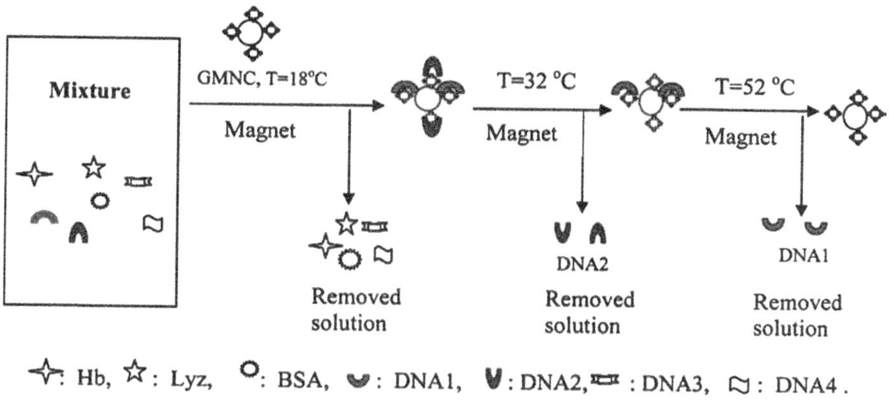

Fig. 5.6 Working principle of the MB-immobilized MNPs for separation and collection of gene products with single-base mismatch discrimination ability (Reprinted with the permission from Ref. [56]. Copyright 2012 American Chemical Society)

Therefore, at 32 °C, single-base mismatched DNA was completely dissociated from the MB-MNP conjugates, and only the complementary target DNA was still captured. When the temperature was further increased to 52 °C, complementary target DNA was also completely dissociated from the conjugates, regenerating them to be used again. Their results showed that the strategy worked perfectly well in buffer solution, as well as cell samples. Highly efficient collection of rare amount of DNA/RNA samples, as low as femtomolar concentrations, was achieved. In addition, the system had excellent ability to differentiate single-base mismatched DNA/RNA samples.

5.6 Conclusion and Outlook

MBs can be non-covalently or covalently immobilized on the surface of bulk solid support matrixes, including microscope glass slides, gold surfaces, and optical fibers. Non-covalent immobilization of MBs on these surfaces is commonly realized by biotin-(strept)avidin interaction, while covalent immobilization is generally based on siloxane-based chemistry. Compared to microscope glass slides and optical fibers, gold surfaces serve as supporting substrates and quenching agents. Other than various bulk solid surfaces, MBs can also be immobilized on the surface of solid nanomaterial support matrixes, taking advantage of the unique properties of nanomaterials. As a result of surface effects, MBs on solid surfaces normally suffer high background and reduced fluorescence enhancements, especially for MBs immobilized on microscope glass slides. Therefore, different immobilization techniques and surface treatment approaches have been tried to improve the performances of surface-immobilized MBs.

Even though the performance of surface-immobilized MBs has been significantly improved by using the newly developed immobilization techniques and surface treatment approaches, further improvement is still desired in the future. Moreover, the use of functional solid supports other than gold surfaces will be an area of great interest. Finally, the reaction thermodynamics and kinetics of surface-immobilized MBs should be carefully and systematically investigated to better understand their behavior on solid surfaces. The more that is known about how MBs behave on solid surfaces, the greater the chance that surface-immobilized MBs with excellent performance can be designed.

References

1. Situma C, Moehring AJ, Noor MAF, Soper SA (2007) Immobilized molecular beacons: a new strategy using UV-activated poly(methyl methacrylate) surfaces to provide large fluorescence sensitivities for reporting on molecular association events. Anal Biochem 363(1):35–45
2. DeLange RJ, Huang T-S (1971) Egg white avidin. J Biol Chem 246(3):698–709
3. Bayer EA, Ben-Hur H, Wilchek M, Meir W, Edward AB (1990) Chapter 8: Isolation and properties of streptavidin. In: Methods in enzymology, vol 184. Academic, New York, pp 80–89
4. Holmberg A, Blomstergren A, Nord O, Lukacs M, Lundeberg J, Uhlén M (2005) The biotin-streptavidin interaction can be reversibly broken using water at elevated temperatures. Electrophoresis 26(3):501–510
5. Angerer L, Davidson N, Murphy W, Lynch D, Attardi G (1976) An electron microscope study of the relative positions of the 4S and ribosomal RNA genes in HeLa cell mitochondrial DNA. Cell 9(1):81–90
6. Diamandis E, Christopoulos T (1991) The biotin-(strept)avidin system: principles and applications in biotechnology. Clin Chem 37(5):625–636
7. Martinez K, Estevez MC, Wu Y, Phillips JA, Medley CD, Tan W (2009) Locked nucleic acid based beacons for surface interaction studies and biosensor development. Anal Chem 81(9):3448–3454
8. Zhang C-Y, Yeh H-C, Kuroki MT, Wang T-H (2005) Single-quantum-dot-based DNA nanosensor. Nat Mater 4(11):826–831
9. Li J, Tan WH, Wang KM, Xiao D, Yang XH, He XX, Tang ZW (2001) Ultrasensitive optical DNA biosensor based on surface immobilization of molecular beacon by a bridge structure. Anal Sci 17(10):1149–1153
10. Fang X, Liu X, Schuster S, Tan W (1999) Designing a novel molecular beacon for surface-immobilized DNA hybridization studies. J Am Chem Soc 121(12):2921–2922
11. Broude NE, Woodward K, Cavallo R, Cantor CR, Englert D (2001) DNA microarrays with stem-loop DNA probes: preparation and applications. Nucleic Acids Res 29(19):e92
12. Guo Z, Guilfoyle RA, Thiel AJ, Wang R, Smith LM (1994) Direct fluorescence analysis of genetic polymorphisms by hybridization with oligonucleotide arrays on glass supports. Nucleic Acids Res 22(24):5456–5465
13. Ramachandran A, Flinchbaugh J, Ayoubi P, Olah GA, Malayer JR (2004) Target discrimination by surface-immobilized molecular beacons designed to detect Francisella tularensis. Biosens Bioelectron 19(7):727–736
14. Osborne MA, Barnes CL, Balasubramanian S, Klenerman D (2001) Probing DNA surface attachment and local environment using single molecule spectroscopy. J Phys Chem B 105(15):3120–3126

15. Zuo X, Yang X, Wang K, Tan W, Li H, Zhou L, Wen J, Zhang H (2006) Improving the performance of immobilized molecular beacons through cleavage. Anal Chim Acta 567(2):173–178
16. Liu H, Wang H, Shi Z, Wang H, Yang C, Silke S, Tan W, Lu Z (2006) TaqMan probe array for quantitative detection of DNA targets. Nucleic Acids Res 34(1):e4
17. Yao G, Tan W (2004) Molecular-beacon-based array for sensitive DNA analysis. Anal Biochem 331(2):216–223
18. Koshkin AA, Nielsen P, Meldgaard M, Rajwanshi VK, Singh SK, Wengel J (1998) LNA (Locked nucleic acid): an RNA mimic forming exceedingly stable LNA:LNA duplexes. J Am Chem Soc 120(50):13252–13253
19. Wang L, Yang CJ, Medley CD, Benner SA, Tan W (2005) Locked nucleic acid molecular beacons. J Am Chem Soc 127(45):15664–15665
20. Wang H, Li J, Liu H, Liu Q, Mei Q, Wang Y, Zhu J, He N, Lu Z (2002) Label-free hybridization detection of a single nucleotide mismatch by immobilization of molecular beacons on an agarose film. Nucleic Acids Res 30(12):e61
21. Wang Y, Wang H, Gao L, Liu H, Lu Z, He N (2005) Polyacrylamide gel film immobilized molecular beacon array for single nucleotide mismatch detection. J Nanosci Nanotechnol 5(4):653–658
22. Prabhakar N, Singh H, Malhotra BD (2008) Nucleic acid immobilized polypyrrole-polyvinylsulphonate film for Mycobacterium tuberculosis detection. Electrochem Commun 10(6):821–826
23. Culha M, Stokes DL, Griffin GD, Vo-Dinh T (2004) Application of a miniature biochip using the molecular beacon probe in breast cancer gene BRCA1 detection. Biosens Bioelectron 19(9):1007–1012
24. Du H, Disney MD, Miller BL, Krauss TD (2003) Hybridization-based unquenching of DNA hairpins on Au surfaces: prototypical "Molecular Beacon" biosensors. J Am Chem Soc 125(14):4012–4013
25. Du H, Strohsahl CM, Camera J, Miller BL, Krauss TD (2005) Sensitivity and specificity of metal surface-immobilized "Molecular Beacon" biosensors. J Am Chem Soc 127(21):7932–7940
26. Lin Z, Strother T, Cai W, Cao X, Smith LM, Hamers RJ (2002) DNA attachment and hybridization at the silicon (100) surface. Langmuir 18(3):788–796
27. Tan W, Fang X, Li J, Liu X (2000) Molecular beacons: a novel DNA probe for nucleic acid and protein studies. Chem Eur J 6(7):1107–1111
28. Sauthier ML, Carroll RL, Gorman CB, Franzen S (2002) Nanoparticle layers assembled through DNA hybridization: characterization and optimization. Langmuir 18(5):1825–1830
29. Jin R, Wu G, Li Z, Mirkin CA, Schatz GC (2003) What controls the melting properties of DNA-linked gold nanoparticle assemblies. J Am Chem Soc 125(6):1643–1654
30. Henke L, Piunno PAE, McClure AC, Krull UJ (1997) Covalent immobilization of single-stranded DNA onto optical fibers using various linkers. Anal Chim Acta 344(3):201–213
31. Abel AP, Weller MG, Duveneck GL, Ehrat M, Widmer HM (1996) Fiber-optic evanescent wave biosensor for the detection of oligonucleotides. Anal Chem 68(17):2905–2912
32. Pilevar S, Davis CC, Portugal F (1998) Tapered optical fiber sensor using near-infrared fluorophores to assay hybridization. Anal Chem 70(10):2031–2037
33. Piunno PAE, Krull UJ, Hudson RHE, Damha MJ, Cohen H (1995) Fiber-optic DNA sensor for fluorometric nucleic acid determination. Anal Chem 67(15):2635–2643
34. Kleinjung F, Bier FF, Warsinke A, Scheller FW (1997) Fibre-optic genosensor for specific determination of femtomolar DNA oligomers. Anal Chim Acta 350(1–2):51–58
35. Liu X, Tan W (1999) A fiber-optic evanescent wave DNA biosensor based on novel molecular beacons. Anal Chem 71(22):5054–5059
36. Hutchinson A (1995) Evanescent wave biosensors. Mol Biotechnol 3(1):47–54
37. Liu X, Farmerie W, Schuster S, Tan W (2000) Molecular beacons for DNA biosensors with micrometer to submicrometer dimensions. Anal Biochem 283(1):56–63

38. Chen T, Shukoor MI, Chen Y, Yuan Q, Zhu Z, Zhao Z, Gulbakan B, Tan W (2011) Aptamer-conjugated nanomaterials for bioanalysis and biotechnology applications. Nanoscale 3(2):546–556
39. Jianrong C, Yuqing M, Nongyue H, Xiaohua W, Sijiao L (2004) Nanotechnology and biosensors. Biotechnol Adv 22(7):505–518
40. Lanone S, Boczkowski J (2006) Biomedical applications and potential health risks of nanomaterials: molecular mechanisms. Curr Mol Med 6(6):651–663
41. Yu D-H, Lu Q, Xie J, Fang C, Chen H-Z (2010) Peptide-conjugated biodegradable nanoparticles as a carrier to target paclitaxel to tumor neovasculature. Biomaterials 31(8):2278–2292
42. Meziani MJ, Lin Y, Sun Y-P (2007) Conjugation of nanomaterials with proteins. In: Nanotechnologies for the life sciences. Wiley-VCH Verlag GmbH & Co. KGaA. doi:10.1002/9783527610419.ntls0007
43. Wang H, Yang R, Yang L, Tan W (2009) Nucleic acid conjugated nanomaterials for enhanced molecular recognition. ACS Nano 3(9):2451–2460
44. Dubertret B, Calame M, Libchaber AJ (2001) Single-mismatch detection using gold-quenched fluorescent oligonucleotides. Nat Biotechnol 19(4):365–370
45. Fang Y-M, Song J, Chen J-S, Li S-B, Zhang L, Chen G-N, Sun J-J (2011) Gold nanoparticles for highly sensitive and selective copper ions sensing-old materials with new tricks. J Mater Chem 21(22):7898–7900
46. Hutter E, Fendler JH (2004) Exploitation of localized surface plasmon resonance. Adv Mater 16(19):1685–1706
47. Mao X, Xu H, Zeng Q, Zeng L, Liu G (2009) Molecular beacon-functionalized gold nanoparticles as probes in dry-reagent strip biosensor for DNA analysis. Chem Commun 21:3065–3067
48. Tyagi S (2009) Imaging intracellular RNA distribution and dynamics in living cells. Nat Meth 6(5):331–338
49. Giljohann DA, Seferos DS, Patel PC, Millstone JE, Rosi NL, Mirkin CA (2007) Oligonucleotide loading determines cellular uptake of DNA-modified gold nanoparticles. Nano Lett 7(12):3818–3821
50. Fan P, Suri AK, Fiala R, Live D, Patel DJ (1996) Molecular recognition in the FMN – RNA aptamer complex. J Mol Biol 258(3):480–500
51. Beer D, Weber J (1972) Photobleaching of organic laser dyes. Optics Commun 5(4):307–309
52. Yu WW, Chang E, Drezek R, Colvin VL (2006) Water-soluble quantum dots for biomedical applications. Biochem Biophys Res Commun 348(3):781–786
53. Aldana J, Wang YA, Peng X (2001) Photochemical instability of CdSe nanocrystals coated by hydrophilic thiols. J Am Chem Soc 123(36):8844–8850
54. Guo W, Li JJ, Wang YA, Peng X (2003) Conjugation chemistry and bioapplications of semiconductor box nanocrystals prepared via dendrimer bridging. Chem Mater 15(16):3125–3133
55. Cady NC, Strickland AD, Batt CA (2007) Optimized linkage and quenching strategies for quantum dot molecular beacons. Mol Cell Probes 21(2):116–124
56. Zhao X, Tapec-Dytioco R, Wang K, Tan W (2003) Collection of trace amounts of DNA/mRNA molecules using genomagnetic nanocapturers. Anal Chem 75(14):3476–3483
57. Santra S, Tapec R, Theodoropoulou N, Dobson J, Hebard A, Tan W (2001) Synthesis and characterization of silica-coated iron oxide nanoparticles in microemulsion: the effect of nonionic surfactants. Langmuir 17(10):2900–2906

Chapter 6
Functional Materials for Signal Amplification of Molecular Beacons

Da Han and Weihong Tan

Abstract Rapid development of bioanalysis and biomedicine requires higher sensitivity and selectivity of MBs. Functional materials are generally characterized as those materials which possess particular native properties. In this chapter, we focus on introducing the substantial progress in the development of using functional materials to amplify the signals of MBs. The amplification effect is generated through improving the target recognition and signal transduction processes of MBs. The signal amplification mechanism, as well as the strengths and weaknesses of different functional materials, such as metal materials, carbon materials, and polymers, is explained in detail to facilitate the applications of MBs in the field of chemistry, medicine, and biology.

6.1 Introduction

To achieve higher sensitivity and selectivity, various molecular engineering strategies have been applied to improve the signal-to-background ratio of MBs. In the past decade, substantial progress in the development of functional materials has led to numerous effective and flexible ways to amplify the signals of MBs. The combination of MBs and functional materials can improve the signals of MBs in the following two ways. First, functional materials can improve the recognition ability of MBs. Functional materials can participate in molecular recognition processes

D. Han (✉)
Department of Chemistry, University of Florida, Gainesville, FL 32611, USA

W. Tan
Department of Biomedical Engineering and Department of Chemistry, Hunan University, Changsha, P.R. China

Department of Chemistry, University of Florida, Gainesville, FL, USA
e-mail: tan@chem.ufl.edu

C.J. Yang and W. Tan (eds.), *Molecular Beacons*, DOI 10.1007/978-3-642-39109-5_6,

and increase the binding affinity between MBs and their cognate targets. This might also be considered a functional synergism in which the combined effect of two or more similar components exceed the sum of the effect of one component used alone [1]. In addition, MBs with functional materials can generate polyvalent effect which subsequently increases the local concentration of the probes, thus improving the recognition ability of MBs. Furthermore, by improving the delivery efficiency of MBs, functional materials also improve their sensing ability inside biological systems. Second, functional materials can boost the signal transduction process of MBs. Using suitable functional materials as reporters or quenchers, the signal-to-background ratio of MBs can be enhanced by several orders of magnitude. Specifically, functional materials can either decrease the background signals in the absence of targets or increase readout signals in the presence of targets. Consequently, from the viewpoint of both target recognition and signal transduction, a combination of MBs and functional materials can lead to amplified signals, as well as improved sensitivity and selectivity. In this chapter, we will focus on functional materials, such as metal nanomaterials, carbon nanomaterials, and polymers, which have been used to realize signal amplification of MBs in the fields of chemistry, medicine, and biology.

6.2 Metal Nanomaterials

Over the past 20 years, nanotechnology has experienced extremely rapid development. Metal nanomaterials of various sizes and shapes were synthesized and characterized with different attracting optical, electric, and magnetic properties. When coupled with MBs, those metal nanomaterials can induce synergistic effect on the signal amplification of MBs by both increasing signal-to-background ratio and improving stability.

6.2.1 Gold Nanomaterials

6.2.1.1 Gold Nanomaterials as Superquenchers

In order to have better signal-to-background ratio, MBs require various methods of refinement. One possible solution is to choose suitable quenchers to effectively suppress the fluorescence signal when MBs are in the “OFF” or closed state. Traditional organic quenchers were found to have two potential problems. One is low quenching efficiency, which means a small signal difference between the ON and OFF states. Another problem is the limited use of organic quenchers that only work with specific dyes. Both defects restrict the application of MBs.

The noble metal gold can be conjugated with various DNA biosensors based on its biocompatibility and easy functionalization with biomolecules; moreover, it possesses some unique optical and electronic properties, such as plasmonic effect and fluorescent quenching ability. The special interaction between the mercapto

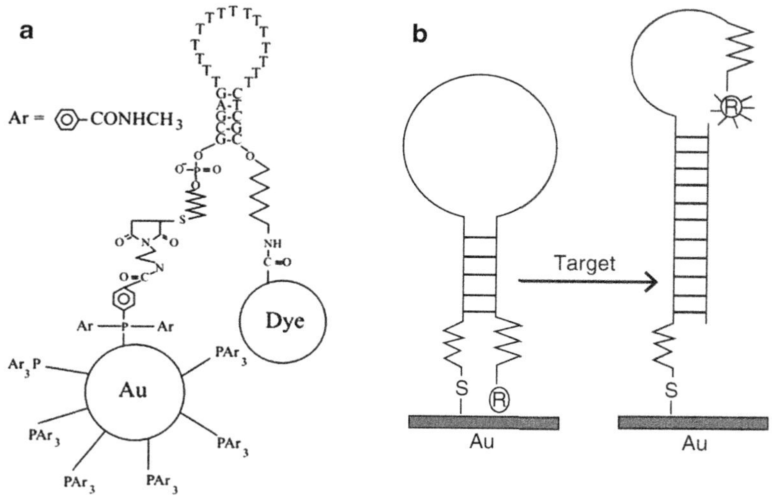

Fig. 6.1 Schematic drawing of gold material-based MBs. (**a**) 1.4 nm AuNP-quenched MBs (**b**) Au nanofilm-immobilized MBs (Reprinted by permission from Macmillan Publishers Ltd: Ref. [3], copyright 2001. Reprinted with the permission from Ref. [6]. Copyright 2005 American Chemical Society)

group and Au atoms facilitates the modification of gold nanomaterials with oligonucleotides [2]. Gold nanoparticles can quench fluorescent dyes based on their efficient absorbance of internal energy, transferred electrons, as well as long-range resonant energy transfer, otherwise known as Förster (fluorescence) resonance energy transfer (FRET), as explained in earlier chapters, from the fluorophores. By applying gold nanoparticles, as the substitution for organic quenchers, to MBs, using either covalent or noncovalent modification, the signal-to-background ratio could be greatly improved. Dubertret et al. constructed the first AuNP-based MB by attaching a fluorophore-tagged stem-loop probe at the surface of small AuNPs (1.4 nm), which led to high quenching efficiency towards a wide range of organic fluorophores (Fig. 6.1a). They found that the sensitivity of these AuNP-based MBs could be increased 100 times, compared with conventional organic dye-quencher beacons. In addition, the ability of MBs to discriminate single-base mismatch is also enhanced by ten times [3]. Later on, larger-sized gold nanoparticles (13 nm) were functionalized with MBs to explore higher quenching efficiency and improve signal output. The higher conjugation density of MBs at the surface of larger-sized AuNPs can also increase the recognition ability of MBs through probe enrichment, which generates signal amplification for target molecules [4]. However, even though larger amounts of MBs could be carried on larger-sized AuNPs, further study exploring these larger AuNPs (50 nm) revealed their unsuitability for amplifying signals by the undesired quenching effect on the "ON" or open state [5].

Besides different sizes, gold materials with various morphologies were also introduced to improve the sensitivities of MBs. Gold films were used to quench the fluorescence of the MBs by Krauss et al. [6]. In their design, (Fig. 6.1b) MBs were immobilized on gold film, which is not only a quencher but also a supporter. The fluorescence intensity of MBs could increase up to 100 times when binding to their targets. In addition to gold film, gold nanowire was demonstrated to exhibit unique enrichment effect and outstanding quenching efficiency for MBs [7]. This research speeded up the development of microarray-based MBs systems.

6.2.1.2 Gold Nanomaterials as Carriers

MBs can be used for intracellular mRNA detection and visualization [8]. Upon intracellular hybridization with the target RNA, MBs are switched to the "ON" state. This event results in fluorescence emission attributable to the intracellular presence of mRNA. However, the applications of MBs in visualizing and detecting intracellular mRNA are limited by the lack of an efficient delivery system. Without an effective carrier, MBs suffer from low uptake efficiency and considerable cellular enzymatic degradation, which lead to a high background signal, as well as a decreased ability to specifically detect target structures inside the cells. To improve this situation, gold nanomaterials have been demonstrated to be an excellent nanocarrier for intracellular delivery since gold nanomaterials (1) have low toxicity and are biocompatible with appropriate surface modifications, (2) can be effectively internalized into cells through endocytosis, and (3) can effectively prevent intracellular enzyme digestion to the surface-functionalized biomolecules, such as DNA or peptides [9]. Also, MBs conjugated with gold nanocarriers were designed to perform more sensitive and selective intracellular target recognition [10]. In this study, a hairpin DNA structure (MBs) consists of a 5′ thiol followed by a 10-base polythymine linker sequence and a stem-loop-stem sequence with a fluorophore at its 3′-end. The loop region is designed to hybridize with specific mRNA sequences of respiratory syncytial virus (RSV) which are in infected mammalian cells (HEp-2). After incubating the AuNP-MBs with live cells, high internalization efficiency and obvious signal differences for infected and uninfected cells were observed from the fluorescent microscopy images (Fig. 6.2). Consequently, in the context of cellular applications, it was demonstrated that the combination of MBs and gold nanomaterials could lead to better spatial intracellular localization and visualization effect on target molecules, thereby increasing the sensitivity of MBs.

6.2.2 *Silica Nanomaterials*

With some unique properties, such as biocompatibility, well-defined morphology, and versatile groups for further biomolecular functionalization, silica nanomaterials have been widely employed in biological applications [11]. Silica nanomaterials have been demonstrated to have unique advantages when combined with biomolecules for bioanalysis [12–14].

Fluorescent silica nanoparticles have emerged as promising fluorescent probes and have attracted widespread interest in biology and medicine [12–14]. Fluorescent

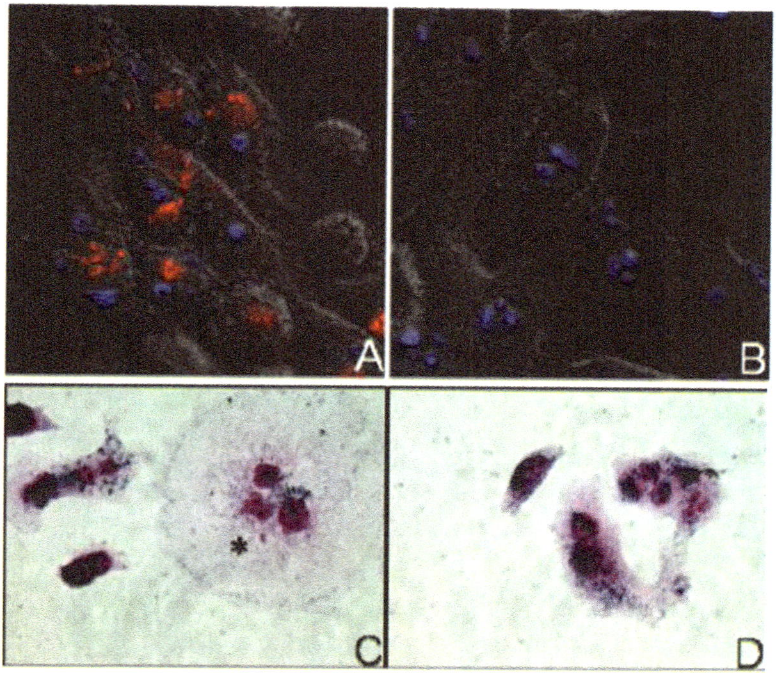

Fig. 6.2 Intracellular uptake and fluorescence emission of AuNP-functionalized MBs. (**a**) RSV-infected cells (**b**) uninfected cells. *Blue*, nuclear counterstaining. Gold nanoparticle immunohistochemical staining indicates that AuNP-MB complexes are internalized within (**c**) RSV-infected HEp-2 cells or (**d**) uninfected cells. Uptake occurs irrespective of cell infection and the intended target of hairpin DNA (Reprinted with the permission from Ref. [10]. Copyright 2010 American Chemical Society.)

silica nanoparticles are fabricated by doping fluorescent dyes into the core part of nanoparticles. The high encapsulation efficiency for organic dyes can be achieved as a result of the intensely porous structures in silica, which subsequently amplify the fluorescence signal approximately 104 times higher compared with single fluorophore-labeled probes [15]. In addition, the fluorescent silica nanoparticles have excellent photostability and water solubility. Consequently, by modifying hybrid fluorescent silica nanoparticles with functional oligonucleotides, the sensitivity of those biomolecular probes can be greatly enhanced. For example, Tris(2, 2′-bipyridyl) dichlororuthenium (II) hexahydrate (RuBpy)-doped silica NPs (60 nm in diameter, each containing >10,000 dye molecules) have been used as highly sensitive and photostable labels in Affymetrix GeneChips technology [12]. Using the RuBpy-doped silica nanoparticles as probes for ssDNA targets, a concentration detection limit of 50 fM was achieved. This work demonstrated the ability of using luminescent NPs as probes for commercial microarray systems, making them less costly, more reproducible, and more sensitive than conventional methods.

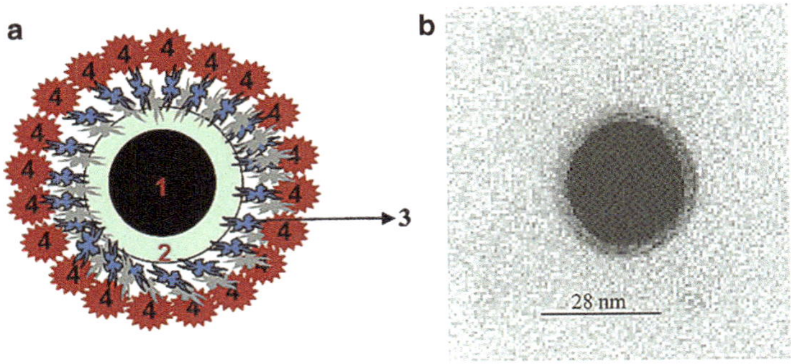

Fig. 6.3 Structure of the magnetic silica nanoparticles. (**a**) Schematic representation of a MB-functionalized core-shell magnetic silica nanoparticle: (1) magnetic nanoparticle; (2) silica layer; (3) biotin-avidin linkage; (4) molecular beacon DNA probe. (**b**) TEM image of a silica-coated magnetic nanoparticle. The diameter is in the range of 28 nm (Reprinted with the permission from Ref. [16]. Copyright 2003 American Chemical Society.)

Magnetic silica nanoparticles represent another important hybrid material which can be used for amplifying signals of MBs. The core of this magnetic nanoparticle is made of iron oxide which can be used for separation. The shell part is silica which allows for the conjugation of various biomolecular probes (Fig. 6.3b). Through magnetic separation, the target molecules can be enriched at the surface of particles and thus improve the sensing ability of MBs. Zhao et al. used this MB-functionalized core-shell magnetic silica nanoparticle complex to collect, separate, and detect trace amount of DNA/RNA (Fig. 6.3a). The highly efficient collection ability and effective signal restoration of this MBs-nanoparticle complex lowered the detection limit of DNA/mRNA samples down to femtomolar (10^{-15} M) concentrations [16].

6.2.3 *Quantum Dots*

Quantum dots (QDs) are a group of semiconductor materials which possess unique photophysical properties [17]. For instance, the high fluorescence quantum yields and stability of QDs improve sensitivity and prolong lifetime in their use as optical labels. Their size-controlled luminescence permits the use of QDs as fluorescence tags for multiplex analysis. Meanwhile, broad absorption bands allow exciting different QDs at a common wavelength simultaneously while imaging. Furthermore, QDs are stable against photobleaching. Compared with traditional organic dyes, including, for example, the family of Rhodamine chemical compounds, semiconductor QDs are approximately 2 orders of magnitude brighter, 100 times stable against photobleaching, and one third as wide by spectral linewidth [18]. More importantly, QDs are water soluble and biocompatible. All those properties make QDs excellent nanomaterials for biosensing applications [19, 20].

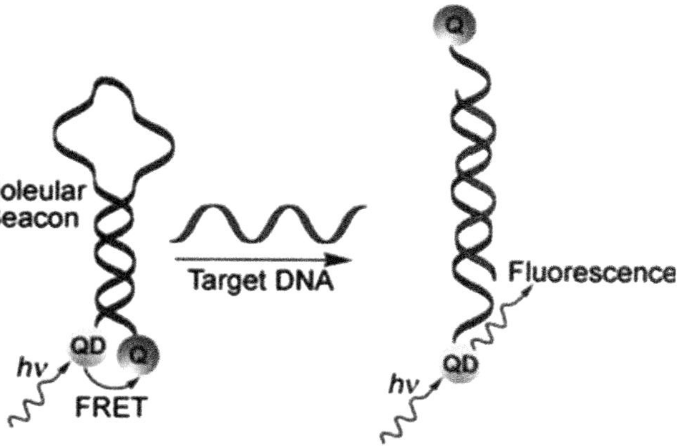

Fig. 6.4 Schematic illustration of QD-based MBs and their sensing principle (Reproduced from Ref. [19] by permission of John Wiley & Sons Ltd.)

QDs are easily modified with biomolecules, such as oligonucleotides and antibodies. Among those molecules, QDs integrated with functional nucleic acids result in a new category of probes for molecular recognition and biosensing. Cady et al. developed several QD-based MBs for DNA-based detection [21]. Compared with traditional MBs with organic dye as fluorophore, QD-based MBs employ QDs as the source of luminescence (Fig. 6.4). The MBs are covalently conjugated to QDs mainly through two methodologies. One is coupling amino linker-modified MBs with carboxyl-functionalized QDs. The other employs the interaction of streptavidin-biotin modified on QDs and MBs, respectively. The streptavidin-biotin pairs were found to have higher FRET efficiencies compared with amine-carboxyl pairs [22]. Various quenchers have been used to suppress the fluorescence of QDs in QD-based MBs. For example, CdSe/ZnS QDs were linked to the 5′-end of a hairpin nucleic acid that included the quencher molecule 4-(4′ dimethylaminophenylazo) benzoic acid (Dabcyl) at the 3′-end. Similarly, CdSe/Zns QDs were conjugated to hairpin DNAs which were linked at the other ends to the Black Hole Quencher-2 (BHQ2) units. Both QDs-quencher pairs were demonstrated to be effective and have long-term stability. They showed high target discrimination upon hybridization with the complementary target DNA, single-base mismatched target DNA (SMT), and nonspecific DNA (NST). A 90 % fluorescence intensity increase was observed in the presence of complementary target DNA relative to the system that included a nonspecific target DNA [23, 24].

Because of their broad absorption bands, QDs can be used for multiplex target detection with different fluorescent emissions excited simultaneously. By coupling different MBs with QDs of different sizes, different target molecules could be recognized at the same time. This strategy greatly improved the throughput of MBs, realizing, in turn, rapid and sensitive detection. Similar to other metal nanomaterial-modified MBs, QDs can also perform as concentrators to amplify the target signals by confining several target molecules in a nanoscale domain. Zhang et al. demonstrated the excellent performance characteristics of a QD-FRET nanosensor for DNA detection with ultrahigh sensitivity, excellent selectivity, and great simplicity [25]. The detection limit of their nanosensing system is 100-fold greater than conventional FRET probe-based assays as monitored by confocal fluorescence spectroscopy. In addition, with strong antiphotobleaching properties, QD-based MBs are stable in complex environment, which means they will be very good candidates for intracellular imaging [25].

6.3 Carbon Materials

6.3.1 Carbon Nanotubes

Carbon nanotubes (NTs) are hollow cylindrical tubes which are made of one or more concentric sheets of sp^2 carbons with diameters ranging from 1 to 100 nm [26]. According to their different structures, NTs can be classified as single-wall carbon nanotubes (SWNTs) and multi-wall carbon nanotubes. With promising mechanical, electrical, and optoelectrical properties, NTs have been termed as "materials of the twenty-first century."

SWNTs are made of one graphene sheet with narrow diameter ranging from 0.4 to 3 nm. It has been reported that semiconducting SWNTs have sharp densities of electronic states at the van Hove singularities, which is a kink or "discontinuity" in the density of states (DOS) of a solid, leading to unique optical properties, such as near-IR absorption and photoluminescence [27]. As a result, SWNTs have been exploited for the applications of photothermal therapy [28, 29] and biological imaging [30, 31]. Recently, SWNTs were demonstrated to be extremely effective quenchers for fluorophores, including organic dyes and fluorophore-labeled biomolecules [32]. Basically the energy-transfer and electron-transfer processes occur between the fluorophores and SWNTs which can deactivate the fluorescence emission process. In particular, SWNTs have an absorption spectrum spanning a wide range of wavelengths (500–900 nm), significantly overlapping the photoluminescence spectra of various fluorophores. This allows FRET to occur. Consequently, if coupled with recognition molecules, SWNTs can be applied as effective quenchers for the detection of various targets.

Single-stranded DNA (ssDNA) has recently been demonstrated to interact noncovalently with SWNTs [32]. ssDNA molecules can wrap around the surface of SWNTs through π-stacking interactions between the nucleotide bases and the carbon rings of SWNTs. This adsorption between ssDNA and SWNTs is strong and therefore effectively quenches the fluorescence if fluorophores are labeled on ssDNA. Absorbance between double-stranded DNA (dsDNA) and SWNTs was reported to be significantly weaker than that of ssDNA by the competing hybridization between nucleotide bases and their complementary bases. Based on

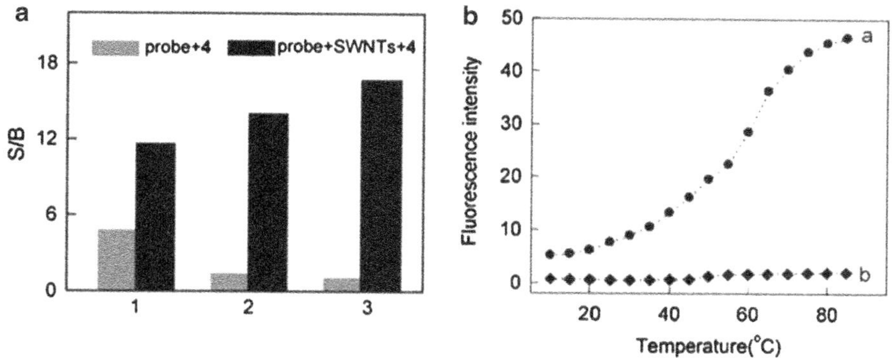

Fig. 6.5 (**a**) Comparisons of the signal-to-background ratio (S/B) of the fluorescent oligonucleotides generated by their perfectly complementary target in the absence (*grey bars*) and presence (*black bars*) of SWNTs: (1) Conventional MBs; (2) One-fluorophore-labeled MBs; (3) One-fluorophore-labeled linear probe. (**b**) Effects of temperature on conventional MBs and SWNT-based MBs. *line a.* conventional MBs; *line b.* SWNT-based MBs (Reprinted with the permission from Ref. [32]. Copyright 2008 American Chemical Society.)

the different binding abilities of ssDNA and dsDNA with SWNTs, a new set of SWNTs-ssDNA complex biosensors was developed, and these biosensors were demonstrated to have better signal-to-noise ratio compared to MBs.

Yang et al. first developed the use of self-assembled quenched complex of fluorescent ssDNA and SWNTs as an efficient substitution of MB that can fluorescently detect single-nucleotide variations in DNA in homogeneous solution [32]. In their design, ssDNA or MB molecules wrap around the individual SWNTs through π-stacking interactions between the nucleotide bases and the carbon rings of SWNTs. As mentioned before, the excellent quenching ability of SWNTs ensures a low background fluorescence signal in the initial state. In this way, only one end of ssDNA or MBs needs to be labeled with a fluorophore, which simplifies the design and synthesis of ssDNA or MBs. In the presence of target molecules, which can hybridize with ssDNA or MBs to form dsDNA, the fluorescence quenching was suppressed, resulting in large fluorescence enhancement from the low affinity between dsDNA and SWNTs. Compared with dual-labeled MBs, the signal-to-background ratio was improved by ten times with the presence of SWNTs (Fig. 6.5a). Moreover, the thermostability of SWNT-MB complexes is superior to that of conventional MBs. Meanwhile, the background fluorescence intensity of SWNT-MB complexes remained stable at temperatures ranging from 15 to 85 °C, while that of regular MBs started to increase at 30 °C (Fig. 6.5b). Consequently, studies involving the SWNT-quenched fluorescent oligonucleotides provide insights into improving the performance, thermostability, and S/Bs of MBs, which are now widely used in many areas of research.

Intracellular mRNA detection is one of the most important applications for MBs. Traditional MBs suffer from such drawbacks as false signals coming from unspecific binding with intracellular proteins and high background signals generated by cellular enzyme digestion. SWNTs can provide an efficient delivery carrier to shuttle oligonucleotides into cells. Furthermore, SWNTs can protect against nuclease digestion or single-strand binding protein interaction of MBs. Wu et al. used the SWNT-based MBs to detect the manganese superoxide dismutase (MnSOD) messenger RNA in living cells [33]. Their results indicated that MB molecules were well protected by SWNTs from cleavage, even after an incubation of 1 h with DNase, an enzyme which can unselectively cleave ssDNA or dsDNA. After incubation of the SWNT-MB complex, the target cells exhibited a high fluorescence signal, while little fluorescence was emitted for the non-target cells. Therefore, SWNT-based MBs with self-delivery capability and intracellular biostability are a significant improvement over traditional MBs for in vivo studies.

In conclusion, SWNTs can realize superior signal amplification for MBs for the following reasons. First, fluorophores can be effectively quenched by SWNTs; as reported, more than 98 % quenching efficiencies were observed. This greatly lowers the background signal and thus increases the S/B ratio of MBs. Second, SWNTs can effectively protect the oligonucleotides against enzymatic digestion, which subsequently improves the performance of MBs in intracellular applications. Finally, SWNTs can lower the false signals generated by nonspecific binding between MBs and ssDNA proteins in a cellular environment.

6.3.2 Graphene Oxide

Graphene is a single atom-thick and two-dimensional carbon material. As an extremely attracting functional material with remarkable electronic, mechanical, and thermal properties, graphene was termed as a "twenty-first-century material" similar to carbon nanotubes. Graphene oxide (GO) is an atomically thin sheet of graphite that has traditionally served as the precursor for graphene. Compared with graphene, GO is insulating and defective which disqualifies this material for certain applications in the view of physicists. However, because of its heterogeneous chemical and electronic structures, along with the fact that it can be processed in aqueous solution, chemists are increasingly drawn to GO [34].

Structurally, GO is an electronically hybrid material that features both conducting π-states from sp^2 carbon sites and a large energy gap (carrier transport gap) between the σ-states of its sp^3-bonded carbons. Meanwhile, the presence of large numbers of oxygen-containing functional groups, such as carboxyl, epoxide, and hydroxyl groups on the GO surface, makes GO extremely hydrophilic and able to be dispersed into single sheets in water and polar organic solvents [34]. Similar to SWNTs, GO can also quench fluorescence generated from dye molecules absorbed on its surface. Combining the two properties of hydrophilicity and fluorescent quenching, GO has been used as a substitute for SWNTs for the design of biosensing platforms [35, 36].

The presence of ionic groups and aromatic domains suggests that GO can interact with biomolecules in a number of ways. Ionic groups, such as O- and

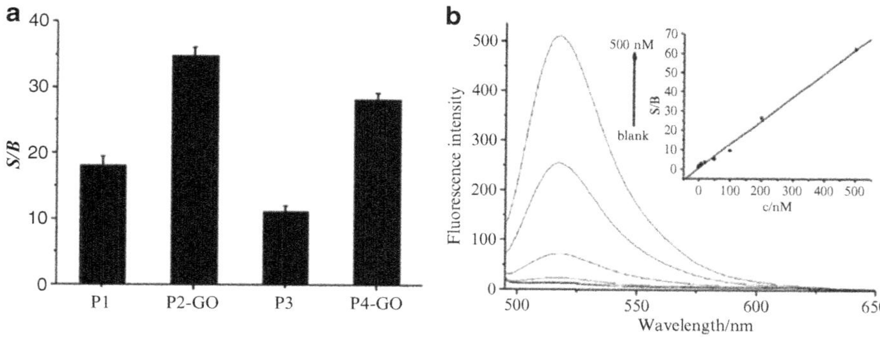

Fig. 6.6 (**a**) Comparisons of the signal-to-background ratio (S/B) generated by a fivefold excess of the DNA target T1. P1: MB with FAM-Dabcyl pairs; P2-GO: FAM-labeled hairpin probe attached on GO. P3: MB with Cy5-Dabcyl pairs; P4-GO: Cy5-labeld hairpin probe attached on GO. The concentrations of P1, P2, P3, and P4 were 50 nm. (**b**) Fluorescence emission spectra of P2–GO (50 nm) in the presence of different concentrations of target T1. *Inset*: signal-to-background ratio (S/B) plotted against the concentration of T1 (Reproduced from Ref. [37] by permission of John Wiley & Sons Ltd.)

COO– that decorate the planes and edges of GO, allowing electrostatic interactions with charged proteins and deoxyribonucleic acid (DNA), while the aromatic scaffold provides a platform for π-stacking and quenching of dyes. In addition, those ionic groups also allow the covalent modification of sensing molecules on the surface of GO. Towards this end, florescence quenching has been used as the basis of GO optical sensors for sensing ssDNA and other biomolecules. The combination of MBs and GO for sensing started in 2009. By absorbing MBs on the GO surface, Lu et al. have demonstrated that the sensitivity can be improved for probing ssDNA targets compared with traditional MBs [37]. MBs labeled with Cy5 and FAM fluorophores were incubated with GO individually. The quenching efficiencies were found to be 99.3 and 99.1 % for those two dyes, respectively, values which are superior to those of any other functional materials mentioned above. Against 250 nM concentration of target molecules, the fluorescence signal increased 35 times for FAM-labeled beacon-GO complex, while that of conventional MBs increased only 18 times (Fig. 6.6a). Those GO-based MBs also possessed other advantages for ssDNA detection, such as sensitive detection (detection of limit: 5 nM) (Fig. 6.6b), ability to discriminate single-base mutations, and fast fluorescence restoration kinetics. Later on, Fan et al. incubated MBs or ssDNA having various fluorophores with GO and used those complexes to simultaneously detect a set of ssDNA targets [38]. Three ssDNA targets were demonstrated to be detected in homogeneous buffer solutions. This strategy, to some extent, amplifies the use of GO-based MBs for multi-color detection.

To explain why GO is proficient at signal amplification, several statements can be made. First, the two-dimensional sheet of GO is equipped with a range of functional groups that can interact in an ionic, covalent, or noncovalent manner so that, in principle, it provides the highest extraction efficiencies of biomolecules per unit area of virtually any material. Second, the exceptional quenching ability of GO significantly lowers the background signals of MBs (even lower than those of SWNT-quenched MBs) even lower than those of SWNT-quenched MBs. Finally, the allowance of multiple target detection improves the throughput of MBs, which, to some extent, also amplifies the final signals.

6.4 Polymers and Other Functional Materials

6.4.1 *Polymers*

Polymers are macromolecules composed of repeating structural units. As functional materials, generation of polymers with well-defined monomer sequences and dispersity, as well as complex functions, are always the most exciting goals for chemical synthesis. Block copolymers, especially those containing biological macromolecules, represent an important step towards achieving this goal [39]. Through combining various macromolecules with different functions, synergistic effects can be achieved with broad applications. DNA itself is among the most extensively studied biomacromolecule. When coupled with polymers, the unique self-recognition properties of DNA, as well as the unique functions of polymers, allow direct application in, for example, drug delivery [40] and sensitive detection [41].

The discovery of some fluorescent polymers prompted their use in sensitive detection with MBs. For instance, poly (phenylene ethynylene) (PPE) is a water-soluble polyelectrolyte with a high quantum yield. A single PPE chain contains several monomers which can generate much higher intensity of fluorescence than conventional organic dyes. At the concentration of 10 nM, it was demonstrated that PPE is about 20 times brighter than Cy3 and six times brighter than Alexa Fluor 488 [42] (Fig. 6.7b). In addition, PPE can be effectively quenched by some organic quenchers, such as dabcyl. With such excellent sensing properties, a single PPE chain was used as a fluorophore with dabcyl as a quencher (Fig. 6.7a). This configuration increased the sensitivity of MBs, enabling them to discriminate single-base mutations. Later on, another PPE-based MBs was developed to broaden the applications of polymer-functionalized MBs for the sensitive detection of target ssDNA [43]. In their design, the combination of twin probe and conjugated polymer (PPE) molecular beacon architectures was realized by attaching two stem-loop structures with terminal quenchers to the two ends of a negatively charged PPE polymer. At target concentrations of 4 μM, the fluorescence signal from the conjugated polymer was two orders of magnitude higher than the signal achieved with a noncomplementary sequence, which greatly amplified the signal of MBs.

Conjugated polyelectrolytes (CPs) are another category of polymers which were coupled with MBs to realize sensitive and selective DNA detection. These

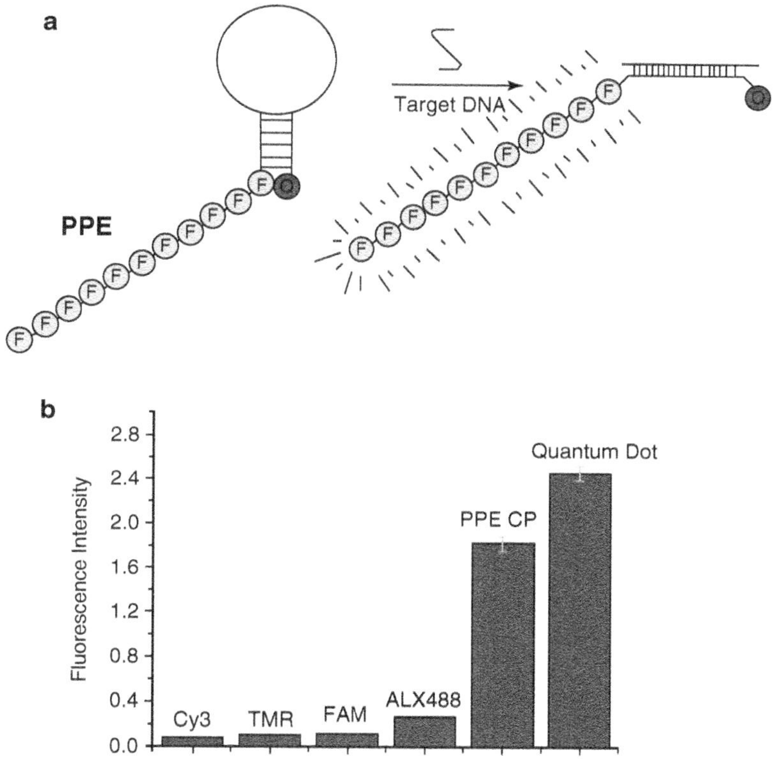

Fig. 6.7 (**a**) An illustration of MB with light-emitting polymer (PPE). (**b**) Comparison of the fluorescence intensity of different fluorophores. The excitation/emission wavelengths for Cy3, TMR, FAM, Alexa Fluor 488, PPE, and quantum dots are 543/560, 557/581, 488/514, 488/515, 440/520, and 400/520 nm, respectively. The concentration of all dyes is 10 nm (Reproduced from Ref. [42] by permission of John Wiley & Sons Ltd.)

conjugated polyelectrolytes can collect and concentrate a large number of absorbing units, which are, in this case, MBs, and then transfer the excitation energy along the backbone to reporter chromophores in MBs, resulting in signal amplification of fluorescence signals [44]. MBs with fluorescein as fluorophore were incubated with a solution containing conjugated polyelectrolyte (PFP-NMe_3^+) and ethidium bromide (EB). The electrostatic interactions between MBs and PFP-NMe_3^+ can generate FRET from PFP-NMe_3^+ to fluorescein. When target DNA was introduced to the system, MBs can be opened to form double-stranded DNA with EB intercalated. EB is a specific intercalator of dsDNA. A two-step FRET then happened, first from PFP-NMe_3^+ to fluorescein and then from fluorescein to EB. By monitoring the emission signal from EB, target DNA could be detected. Based on their results, a single-nucleotide mismatch could be clearly detected in target DNA in 150 s.

6.4.2 *Other Functional Materials*

In addition to polymers, some other molecules were demonstrated to improve the sensitivity of MBs. Asanuma et al. developed an artificial dye molecule called threoninol nucleotide which can intercalate into the stem of MBs [45]. The threoninol nucleotide can be further modified with fluorescent perylene and present as pseudo base pairs in MBs. Their data showed that this artificial dye molecule can suppress the background signal of MBs effectively, resulting in sensitive discrimination of single-nucleotide mismatch. Soper et al. conjugated the novel near-IR dye metallophthalocyanines (Pcs) on both the 5′-end and 3′-end of MBs [46]. The Pcs can self-quench each other while in closed state but will generate near-IR fluorescence when opened by target DNA. These properties of Pcs can increase the photostability of reporting fluorophores in MBs to prolong observation time in imaging applications and conducting assays in the near-IR to avoid the otherwise significant autofluorescence.

6.5 Conclusions

As an effective approach to optimize both target recognition and signal transduction processes, the use of functional materials to improve the performance of MBs has become an effective research model to achieve higher sensitivity and selectivity in bioanalysis and biomedicine. However, most of the advances for functional material-MB combinations, as presented in this chapter, are in vitro studies, with some problems still remaining for the use of these complexes in in vitro studies. For instance, the potential toxicity of a large amount of functional materials to the human body is still not clear, which slows down the clinical development of functional material-based MBs. Moreover, the performance of some functional materials might be less effective in a complex environment, especially in living systems, thus depressing the effectiveness of MBs in vitro. To overcome these challenges, more studies need to be done to understand the properties of functional materials, including how those materials interact with MBs and what the materials do inside the bodies. By successfully solving those problems, functional material coupled with MBs will expand their scope of applications.

References

1. Kim Y, Cao Z, Tan W (2008) Molecular assembly for high-performance bivalent nucleic acid inhibitor. Proc Natl Acad Sci U S A 105(15):5664–5669
2. Mirkin CA, Letsinger RL, Mucic RC, Storhoff JJ (1996) A DNA-based method for rationally assembling nanoparticles into macroscopic materials. Nature 382(6592):607–609
3. Dubertret B, Calame M, Libchaber AJ (2001) Single-mismatch detection using gold-quenched fluorescent oligonucleotides. Nat Biotechnol 19(4):365–370

4. Song S, Liang Z, Zhang J, Wang L, Li G, Fan C (2009) Gold-nanoparticle-based multicolor nanobeacons for sequence-specific DNA analysis. Angew Chem Int Ed Engl 48(46):8670–8674
5. Fan C, Wang S, Hong JW, Bazan GC, Plaxco KW, Heeger AJ (2003) Beyond superquenching: hyper-efficient energy transfer from conjugated polymers to gold nanoparticles. Proc Natl Acad Sci U S A 100(11):6297–6301
6. Du H, Strohsahl CM, Camera J, Miller BL, Krauss TD (2005) Sensitivity and specificity of metal surface-immobilized "molecular beacon" biosensors. J Am Chem Soc 127(21):7932–7940
7. Cederquist KB, Keating CD (2010) Hybridization efficiency of molecular beacons bound to gold nanowires: effect of surface coverage and target length. Langmuir 26(23):18273–18280
8. Bratu DP, Cha BJ, Mhlanga MM, Kramer FR, Tyagi S (2003) Visualizing the distribution and transport of mRNAs in living cells. Proc Natl Acad Sci U S A 100(23):13308–13313
9. Giljohann DA, Seferos DS, Daniel WL, Massich MD, Patel PC, Mirkin CA (2010) Gold nanoparticles for biology and medicine. Angew Chem Int Ed Engl 49(19):3280–3294
10. Jayagopal A, Halfpenny KC, Perez JW, Wright DW (2010) Hairpin DNA-functionalized gold colloids for the imaging of mRNA in live cells. J Am Chem Soc 132(28):9789–9796
11. Wang L, Estevez MC, O'Donoghue M, Tan W (2008) Fluorophore-free luminescent organosilica nanoparticles. Langmuir 24(5):1635–1639
12. Wang L, Lofton C, Popp M, Tan W (2007) Using luminescent nanoparticles as staining probes for Affymetrix GeneChips. Bioconjug Chem 18(3):610–613
13. Wang L, Zhao W, O'Donoghue MB, Tan W (2007) Fluorescent nanoparticles for multiplexed bacteria monitoring. Bioconjug Chem 18(2):297–301
14. Wang L, Yang C, Tan W (2005) Dual-luminophore-doped silica nanoparticles for multiplexed signaling. Nano Lett 5(1):37–43
15. Zhao X, Hilliard LR, Mechery SJ, Wang Y, Bagwe RP, Jin S, Tan W (2004) A rapid bioassay for single bacterial cell quantitation using bioconjugated nanoparticles. Proc Natl Acad Sci U S A 101(42):15027–15032
16. Zhao X, Tapec-Dytioco R, Wang K, Tan W (2003) Collection of trace amounts of DNA/mRNA molecules using genomagnetic nanocapturers. Anal Chem 75(14):3476–3483
17. Bruchez M Jr, Moronne M, Gin P, Weiss S, Alivisatos AP (1998) Semiconductor nanocrystals as fluorescent biological labels. Science 281(5385):2013–2016
18. Chan WC, Nie S (1998) Quantum dot bioconjugates for ultrasensitive nonisotopic detection. Science 281(5385):2016–2018
19. Gill R, Zayats M, Willner I (2008) Semiconductor quantum dots for bioanalysis. Angew Chem Int Ed Engl 47(40):7602–7625
20. Michalet X, Pinaud FF, Bentolila LA, Tsay JM, Doose S, Li JJ, Sundaresan G, Wu AM, Gambhir SS, Weiss S (2005) Quantum dots for live cells, in vitro imaging, and diagnostics. Science 307(5709):538–544
21. Cady NC, Strickland AD, Batt CA (2007) Optimized linkage and quenching strategies for quantum dot molecular beacons. Mol Cell Probes 21(2):116–124
22. Cady NC (2009) Quantum dot molecular beacons for DNA detection. Methods Mol Biol 544:367–379
23. Ozkan M, Kim JH, Morikis D (2004) Adaptation of inorganic quantum dots for stable molecular beacons. Sensors Actuators B-Chem 102(2):315–319
24. Kim JH, Stephens JP, Morikis D, Ozkan M (2004) Hybrid inorganic-organic molecular beacons. Sensor Lett 2(2):85–90
25. Zhang CY, Yeh HC, Kuroki MT, Wang TH (2005) Single-quantum-dot-based DNA nanosensor. Nat Mater 4(11):826–831
26. Britz DA, Khlobystov AN (2006) Noncovalent interactions of molecules with single walled carbon nanotubes. Chem Soc Rev 35(7):637–659
27. Tans SJ, Devoret MH, Dai HJ, Thess A, Smalley RE, Geerligs LJ, Dekker C (1997) Individual single-wall carbon nanotubes as quantum wires. Nature 386(6624):474–477

28. Dai HJ, Kam NWS, O'Connell M, Wisdom JA (2005) Carbon nanotubes as multifunctional biological transporters and near-infrared agents for selective cancer cell destruction. Proc Natl Acad Sci U S A 102(33):11600–11605
29. Vitetta ES, Chakravarty P, Marches R, Zimmerman NS, Swafford ADE, Bajaj P, Musselman IH, Pantano P, Draper RK (2008) Thermal ablation of tumor cells with anti body-functionalized single-walled carbon nanotubes. Proc Natl Acad Sci U S A 105(25):8697–8702
30. Strano MS, Graff RA, Swanson JP, Barone PW, Baik S, Heller DA (2005) Achieving individual-nanotube dispersion at high loading in single-walled carbon nanotube composites. Adv Mater 17(8):980–984
31. Gambhir SS, De La Zerda A, Zavaleta C, Keren S, Vaithilingam S, Bodapati S, Liu Z, Levi J, Smith BR, Ma TJ, Oralkan O, Cheng Z, Chen XY, Dai HJ, Khuri-Yakub BT (2008) Carbon nanotubes as photoacoustic molecular imaging agents in living mice. Nat Nanotechnol 3(9):557–562
32. Yang R, Jin J, Chen Y, Shao N, Kang H, Xiao Z, Tang Z, Wu Y, Zhu Z, Tan W (2008) Carbon nanotube-quenched fluorescent oligonucleotides: probes that fluoresce upon hybridization. J Am Chem Soc 130(26):8351–8358
33. Wu Y, Phillips JA, Liu H, Yang R, Tan W (2008) Carbon nanotubes protect DNA strands during cellular delivery. ACS Nano 2(10):2023–2028
34. Loh KP, Bao Q, Eda G, Chhowalla M (2010) Graphene oxide as a chemically tunable platform for optical applications. Nat Chem 2(12):1015–1024
35. Lu CH, Yang HH, Zhu CL, Chen X, Chen GN (2009) A graphene platform for sensing biomolecules. Angew Chem Int Ed Engl 48(26):4785–4787
36. Lu CH, Li J, Lin MH, Wang YW, Yang HH, Chen X, Chen GN (2010) Amplified aptamer-based assay through catalytic recycling of the analyte. Angew Chem Int Ed Engl 49(45):8454–8457
37. Lu CH, Li J, Liu JJ, Yang HH, Chen X, Chen GN (2010) Increasing the sensitivity and single-base mismatch selectivity of the molecular beacon using graphene oxide as the "nanoquencher". Chemistry 16(16):4889–4894
38. He SJ, Song B, Li D, Zhu CF, Qi WP, Wen YQ, Wang LH, Song SP, Fang HP, Fan CH (2010) A graphene nanoprobe for rapid, sensitive, and multicolor fluorescent DNA analysis. Adv Funct Mater 20(3):453–459
39. Kwak M, Herrmann A (2010) Nucleic acid/organic polymer hybrid materials: synthesis, superstructures, and applications. Angew Chem Int Ed Engl 49(46):8574–8587
40. Zhu Z, Wu C, Liu H, Zou Y, Zhang X, Kang H, Yang CJ, Tan W (2010) An aptamer cross-linked hydrogel as a colorimetric platform for visual detection. Angew Chem Int Ed Engl 49(6):1052–1056
41. Feng X, Liu L, Yang Q, Wang S (2011) Dual-amplified sensitive DNA detection based on conjugated polymers and recyclable autocatalytic hybridization of DNA. Chem Commun (Camb) 47(20):5783–5785
42. Yang CJ, Pinto M, Schanze K, Tan W (2005) Direct synthesis of an oligonucleotide-poly(phenylene ethynylene) conjugate with a precise one-to-one molecular ratio. Angew Chem Int Ed Engl 44(17):2572–2576
43. Kim J, Lee K, Povlich LK (2007) Label-free and self-signal amplifying molecular DNA sensors based on bioconjugated polyelectrolytes. Adv Funct Mater 17(14):2580–2587
44. Feng X, Duan X, Liu L, An L, Feng F, Wang S (2008) Cationic conjugated polyelectrolyte/molecular beacon complex for sensitive, sequence-specific, real-time DNA detection. Langmuir 24(21):12138–12141
45. Asanuma H, Kashida H, Takatsu T, Fujii T, Sekiguchi K, Liang XG, Niwa K, Takase T, Yoshida Y (2009) In-stem molecular beacon containing a pseudo base pair of threoninol nucleotides for the removal of background emission. Angew Chem Int Ed 48(38):7044–7047
46. Soper SA, Nesterova IV, Erdem SS, Pakhomov S, Hammer RP (2009) Phthalocyanine dimerization-based molecular beacons using near-IR fluorescence. J Am Chem Soc 131(7):2432–2433

Chapter 7
Engineering Molecular Beacons for Advanced Applications

Mingxu You, Chaoyong James Yang, and Weihong Tan

Abstract As nucleic acid-based probes, molecular beacons (MBs) have shown promising applications in genetic screening, biosensor development, and single-nucleotide polymorphism detection. The chemical nature of nucleic acids also allows easy synthesis and modification of MBs. As a result, there have been various feasible ways to engineer MBs with advanced bioavailability (e.g., stability, sensitivity, and selectivity in complex environments), regulating ability, and multi-functional properties. In this chapter, recent advances in rational design and novel functionalization of molecular beacons are described. The chapter is divided into three parts covering engineering efforts in the loop region, the stem region, and the reporting region of MBs, respectively. Rational engineering pathways to boost the performance of MBs, expand the ability of MBs for more nongene target detection, and equip them with adjunctive functions are investigated. The broad spectrum of MB engineering and applications is paving the way for the future evolution of bioanalytical and biomedical developments.

M. You (✉)
Department of Chemistry, University of Florida, Gainesville, FL 32611-7200, USA
e-mail: you@chem.ufl.edu

C.J. Yang
Department of Chemical Biology, College of Chemistry and Chemical Engineering, Xiamen University, Xiamen, P.R. China
e-mail: cyyang@xmu.edu.cn

W. Tan
Department of Biomedical Engineering and Department of Chemistry, Hunan University, Changsha, P.R. China

Department of Chemistry, University of Florida, Gainesville, FL, USA
e-mail: tan@chem.ufl.edu

C.J. Yang and W. Tan (eds.), *Molecular Beacons*, DOI 10.1007/978-3-642-39109-5_7,

7.1 Introduction

While molecular beacons (MBs) have been engineered for multiple applications since their discovery in 1996, many challenges and opportunities lie ahead to optimize their performance and extend their functions, particularly for intracellular applications. Current molecular engineering of MBs has focused on generating MB probes with high sensitivity, eliminating false-positive signals, and enhancing intracellular stability and development of regulatable probes for biomedical applications.

For target sequences with a very low copy number, such as found in a single living cell, *the sensitivity of conventional MBs is often insufficient for adequate detection.* As a result, most applications of MBs have been limited to abundant or stimulated gene target reporting [1, 2]. Since the sensitivity of MBs depends on the signal-to-background ratio in the presence of the target, it could be improved by optimizing the reporting system to enhance either quenching efficiency (closed state) or fluorescence intensity (open state). Some nonfluorescent signals could also be employed to develop sensitive probes. Examples demonstrated so far include phosphorescence, chemiluminescence, or employing electrochemical and absorbance signals. Other than reporter modification, proper choice of stem and loop regions can also be crucial, as efficient hybridization with the target will enhance sensitivity, while close proximity of the dyes in the closed state will reduce background signal.

The bioavailability of MBs, especially for in vivo applications, is hindered by their susceptibility to nuclease-mediated digestion and the generation of false-positive signals. Long-term, real-time gene monitoring in living cells using natural nucleotides-based MBs is almost impossible, as MBs were shown to degrade inside cells within around 45 min [3]. As a result, *nuclease-resistant MBs must be generated and employed for intracellular measurements.* The nuclease-resistant MBs developed by far mostly employ some nonnatural nucleotides within the stem and loop regions. The most widely used examples include 2′-*O*-methyl derivatives [4, 5] and MBs based on locked nucleic acids (LNA) [3, 6]. *Other issues hindering bioavailability of MBs are nontarget gene binding and nonspecific protein recognition*, both of which lead to the generation of false-positive signals and inhibit the availability of MBs toward their target genes. To address these problems, homo-MBs [7], reverse-stem-MBs [8], and some nonnatural nucleotide MBs [9, 10] have been engineered, and the detailed mechanisms will be discussed further below.

For molecular engineering of MBs, some novel areas of interest involve *the design of regulatable MBs and MBs capable of performing diverse functions.* The intercalation of chemical or physical triggers has assisted chemists and biologists to fine-tune the functions of MBs or regain control over molecules, even in complex environments. Given the ease of nucleic acid (especially DNA) synthesis and the versatility of modification methodologies, MBs are also being explored for some advanced biological applications. For example, in addition to their function as signaling agents, more and more studies have demonstrated the therapeutic potential of MBs, e.g., in photodynamic therapy (PDT) for targeted cancer treatment [11]. Thus, the rational engineering of nucleic acid MBs with multiple functions is interesting and meaningful, considering the efficient conversion of recognition events into physically detectable signals as well as therapeutic outputs. Further development in this direction could make use of their full potential to meet various requirements of biological systems.

In this chapter, the current state of MB engineering methodologies in order to optimize and extend their performance will be discussed. The chapter is divided into three parts covering engineering efforts in the loop region, the stem region, and the reporting region of MBs, respectively. The reader is reminded, however, that even though different regions are considered separately here, different parts of the MB often work together, and simultaneous and harmonious modifications are often required to fulfill the desired engineering outcome.

7.2 Loop Region Engineering

MBs comprise a target-binding region (loop) flanked by two complementary stem sequences, which are normally unrelated to the targeting sequence. Accordingly, the thermodynamic and kinetic properties of MBs are, to a large extent, related to the favorability of the loop-target hybridization event. As mentioned above, to avoid problems associated with nuclease-mediated degradation, many chemically modified nonnatural nucleotides have been developed, including 2′-*O*-methyl derivatives [7, 8], phosphorothioate derivatives [12, 13], locked nucleic acids (LNA) [9, 10], and peptide nucleic acids (PNA) [14–16]. These modifications are summarized in Fig. 7.1.

One of the earliest attempts to improve the bioavailability of MBs involved 2′-*O*-methyl oligoribonucleotides, which offer good nuclease resistance in living cells. In addition, these artificial nucleotides may also be desirable for hairpin structure

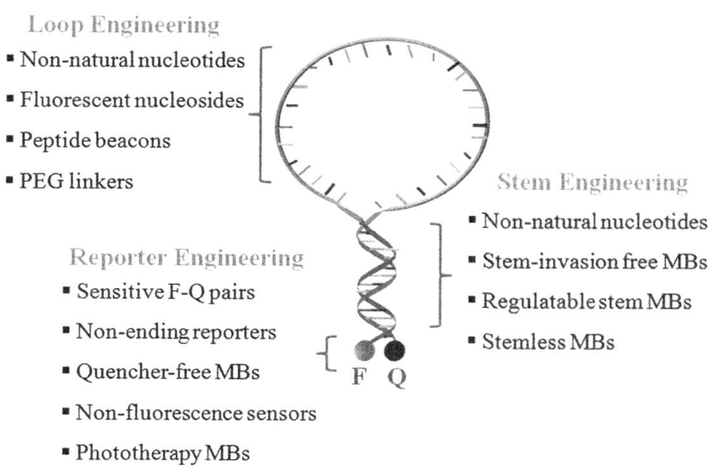

Fig. 7.1 Molecular beacon structures and engineering methodologies for different parts of MBs

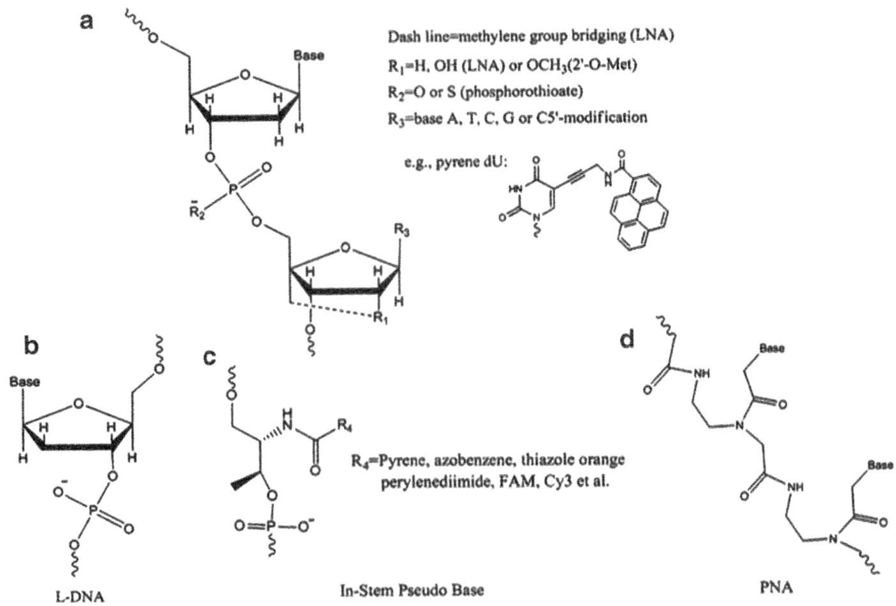

Fig. 7.2 Commonly used nucleotide modifications in MB structures

modification by eliminating the RNA degradation from RNase H, which would otherwise occur upon hybridization of DNA probe with an RNA target in living cells [7]. However, the strong 2′-*O*-methyl nucleotide hybridization can decrease the kinetics of the sensing event. In one interesting study, a chimeric beacon, with a 2′-*O*-methyl loop domain and a 2′-deoxy stem, was used to address this issue of slow hybridization rate in cell studies. Nonetheless, the applications of these 2′-*O*-methyl-modified MBs are still limited, since they can generate false-positive signals in living cells. For example, in one study, nearly 50 % of the MBs opened within a 24-h incubation period with cells, even though no target was present [17]. Rapid nuclear sequestration was the suspected reason, since these beacons would localize in the nucleus and lysosomal compartments after several hours of incubation. In contrast, by linking to a moiety capable of escaping the nucleus (e.g., transfer RNA (tRNA) [18], quantum dots (QDs), or NeutrAvidin proteins [19]), the MBs could be retained within the cytoplasm, significantly reducing the occurrence of false-positive signals. Phosphorothioate internucleotide linkages can also be used to improve duplex stability and nuclease resistance. Combined with 2′-*O*-methyl modification, the derived MBs allowed real-time monitoring of the cell-to-cell spreading of viral infection and did so with decreased false-positive signals [16]. However, by using phosphorothioate derivatives, some studies have shown that backbone modification can occasionally result in toxicity; therefore, these applications have not been fully developed [20] (Fig. 7.2).

Some modified MB structures also employ locked nucleic acid (LNA) bases, which contain a methylene bridge connecting the 2′-oxygen of ribose with the

4′-carbon. LNA bases have advantages over natural nucleic acids (DNA) for MB construction, as a result of the high target binding affinity, excellent base mismatch discrimination capability, and greatly improved nuclease resistance [21]. Duplexes involving LNA, hybridized to either DNA or RNA, display a large increase in melting temperature, ranging from +3.0 to +9.6 °C per LNA modification, compared to the corresponding unmodified polynucleotide. The selectivity of MBs can be further enhanced by using an LNA backbone, which is very crucial in single-nucleotide polymorphism (SNP) studies. However, it is also worth noting that the hybridization kinetics of LNA-MBs can be relatively slow. Similar to 2′-*O*-methyl/2′-deoxy chimeric MBs, DNA/LNA chimeric MBs could also potentially solve this problem. Yang et al. studied the effect of DNA/LNA ratio on the thermodynamics, hybridization kinetics, and enzymatic resistance of MBs [3]. It was found that the hybridization rates of LNA-MBs were significantly improved by lowering the DNA/LNA ratio in the probe. A shared-stem MB with a 4-base-pair stem and alternating DNA/LNA bases is desirable for intracellular applications, as it ensures reasonable hybridization rates, reduces protein binding, and resists nuclease degradation for both target and probes. Furthermore, this type of DNA/LNA chimeric MB [22] was tested with MDA-MB-231 cancer cells to monitor mRNA expression levels in real time for 5–24 h, and it showed extremely high bioavailability. Even after 24 h inside living cancer cells, these LNA-MBs were still functional.

Various groups have also investigated neutral peptide nucleic acids (PNAs) as a scaffold for MBs. With the absence of repeating charges, PNAs are believed to escape degradation by nucleases, and, similar to LNA, PNA-RNA conjugates were normally more stable compared to DNA-RNA or RNA-RNA duplexes [18]. It seems that PNAs may be desirable for MB modification. However, the neutral backbone also creates some undesirable properties, such as the well-documented propensity to self-aggregate [23]. Since the environmental conditions inside living cells do not allow optimum solubility of PNAs, intracellular applications of PNAs are still quite limited. Moreover, since the physical properties of PNAs can change substantially and unpredictably with only small changes in sequence [24], PNA-based MBs might be relatively difficult to manipulate.

The loop domain of MBs is typically used as the target-binding region; however, the generality of the MB approach has always been limited to the detection of oligonucleotide-binding targets. Partially inspired by PNA-MBs, polypeptide-based MB analogues have also been developed to extend the scope of application for MBs [25–28]. Such peptide-based beacons could report the binding (or dissociation) of protein targets using fluorescence intensity changes. However, the development of such peptide beacons has been hampered because stable stem-loop structures are not formed by pure polypeptides. The Seitz group synthesized some hairpin peptide beacons (HPB) [29], which consisted of a central, target-specific peptide sequence (loop) flanked by two PNA arm segments (stem) (Fig. 7.3a). These HPBs featured a stable hairpin structure that could signal the presence of the target protein by an 8- to 11-fold fluorescence increase, much larger than the signal enhancement of traditional peptide beacons. Through the incorporation of peptides with nucleic acids, these examples show that the biological properties of both moieties can be extended and optimized.

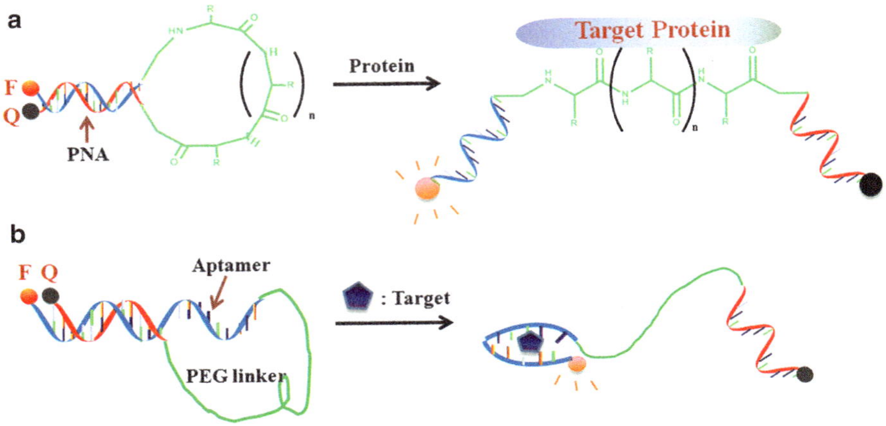

Fig. 7.3 Loop region engineering: (**a**) PNA-stem, peptide-loop beacon for protein targeting; (**b**) aptamer-switch probes for nongene target recognition

Instead of functioning as a target-binding region, the loop domain of MBs can also be simply used as the linker. In one study [30], a polyethylene glycol (PEG) linker was used to connect an aptamer and a short DNA sequence complementary to part of the aptamer. By utilizing the strong and fast intramolecular interactions, such aptamer-switch beacon probes could be used to detect target molecules, such as ATP or human α-thrombin, with high sensitivity and prompt response (Fig. 7.3b). As demonstrated by all of these examples, advances in engineering the loop regions will both increase the bioavailability of the hairpin structures and broaden the potential targets and functions of MBs.

7.3 Stem Region Engineering

The stem region of an MB functions as a "lock" to maintain the closed hairpin structure in the absence of target. Upon hybridization between the loop and its complementary target, the MB undergoes a conformational change from hairpin to linear structure, resulting in increased fluorescence. However, false-positive signals can arise from unwanted opening of the hairpin, mainly from thermal fluctuations and stem invasions. Thermal fluctuations include conformational changes between the hairpin and other thermodynamically possible nonhairpin structures by unwanted intramolecular hybridization between the stem and part of the loop. Stem invasion refers to undesired intermolecular interactions between stem sequences and their complementary sequences, while, ideally, molecular recognition events only occur between the loop region and its complementary target. These problems with the stem region can result in a high background signal and poor analytical selectivity. In

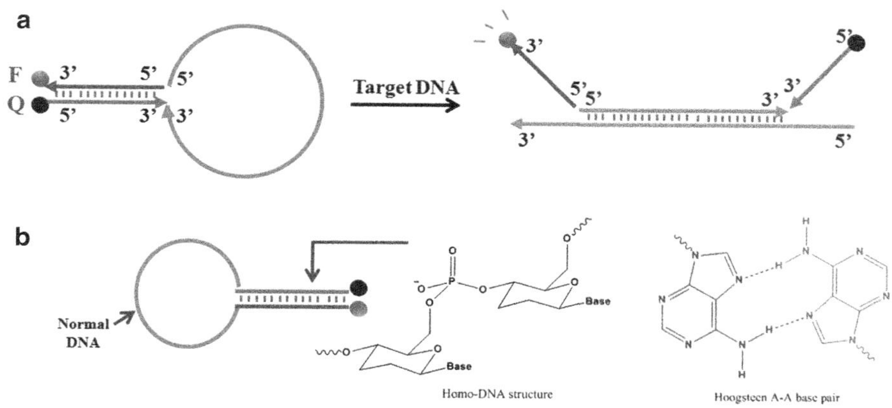

Fig. 7.4 Stem region engineering to avoid stem invasion: (**a**) Hairpin inversion probes with 3′–3′ and 5′–5′ junctions; (**b**) homo-stem MB with Hoogsteen A-A base pairs

order to circumvent these hindrances, various stem modification methods have been developed, with the intention of maintaining stem-stem hybridization while at the same time inhibiting stem-loop and stem-target interactions.

An early attempt in this direction incorporated inversion linkages into the probe backbone, usually by 3′–3′ and 5′–5′ junctions, resulting in the sequence polarity of stem arms opposite to that of the flanking target sequences (e.g., $3' \rightarrow 5'$ stem and $5' \rightarrow 3'$ loop, as shown in Fig. 7.4a) [12]. The so-called hairpin inversion probe (HIP) precludes potential interactions between the stem arms and the flanking target sequences; moreover, the loop sequence is less likely to influence the stability of the stem sequence, due to the discontinuity in the nucleic acid array. Although they demonstrated some success, the usefulness of these MBs is limited. On one hand, stem participation is still not principally excluded, and on the other hand, a longer stem is required, causing sticky-end-pairing problems and complicated syntheses. Shortly after this study, Crey-Desbiolles et al. developed MBs with a homo-DNA stem (Fig. 7.4b) [11]. An orthogonal DNA analog that does not hybridize with natural nucleic acids was used for stem construction. These homo-DNAs could form stable, antiparallel duplexes with themselves and could be easily prepared by standard automated phosphoramidite chemistry for DNA synthesis. By employing nonnatural nucleic acids, these MBs could theoretically be superior to hairpin inversion probes in avoiding stem invasion and thermal fluctuation.

Using a similar strategy, Sheng et al. incorporated an artificially expanded genetic information system, termed A_{EGIS} [10], into the stem of MBs, thereby eliminating unwanted stem invasion at lower cost [14]. In another meaningful trial to optimize the stem of MBs, Kim et al. created an MB with a nonnatural enantiomeric DNA (L-DNA)-incorporated stem domain and natural D-DNA-incorporated loop domain (Fig. 7.2) [9]. L-DNA is attractive in this application because it does not interact with D-DNA by the chirality difference. Moreover, it is quite simple to incorporate

L-DNA into other nonstandard bases, such as 2′-*O*-methyl derivatives. Taking advantage of these features, the L-DNA stem strategy has been demonstrated as a simple, direct, and effective way for optimizing the stem region of MBs with desired sensitivity and stability.

Once the stem is formed, however, there is no means to alter its selectivity and sensitivity, in order to meet some specific needs, posing a key engineering obstacle. One way of overcoming this problem is the development of regulatable MBs. The Seitz group, for instance, developed a triplex MB [29] to enable precise modulation of the conformational constraint. Instead of relying only on inflexible traditional Watson-Crick hydrogen bonding between natural DNA base pairs, a PNA oligonucleotide was designed to bind with two poly-(T) arms in the stem region. Through Watson-Crick hydrogen bonding and Hoogsteen base pairing [31] (Fig. 7.5a), the PNA oligonucleotide allows tuning of the molecular beacon's properties. As expected, melting temperatures of the triplex stem MB increased with increasing PNA strand lengths. The additional opportunity for incorporating more fluorophores or quenchers into the end reporting system could be another merit of these triplex MBs for improved probe sensitivity.

Taking advantage of metal-dependent pairing of two nucleobases (T-Hg^{2+}-T), Yang et al. developed an MB whose target-binding properties could be controlled by Hg^{2+} ions [32], which could selectively bind between two thymine (T) bases and promote these T-T mismatches to form stable T-Hg^{2+}-T base pairs (Fig. 7.5b). At low Hg^{2+} concentration, rapid kinetic response of the MB was achieved; in contrast, decreased background fluorescence, but reduced hybridization rate, was observed at high Hg^{2+} concentration. The Hg^{2+} concentration could be adjusted to make the MB work properly at different temperatures with the desired kinetic response and selectivity. Moreover, these Hg^{2+}-controlled MBs resist both nuclease digestion and single-stranded binding protein (SSB), making them desirable for sensor development. As an example, Lin et al. have used this type of MB to detect single-nucleotide polymorphisms with high sensitivity and selectivity [33].

Another method uses cation concentrations to regulate the properties of MBs and involves G-quadruplex motif incorporation into the stem region (Fig. 7.5c) [34]. Both monovalent and divalent cations (e.g., K^+, Na^+ and Mg^{2+}) can be used in this type of MB. Mismatch discrimination could be achieved at lower temperatures and over a larger temperature range compared to MBs based solely on duplex formation. An interesting application of this type of quadruplex-MB is the recently developed DNAzyme MB [35]. Two separate oligonucleotides that self-assemble into a G-quadruplex structure with hemin were employed as the catalytic unit for generating amplified signals. Without terminal dye modification, this new colorimetric DNA detection system could achieve a detection limit of 1 pM, 3–5 orders of magnitude lower than that of normal DNAzyme MB detection methods. However, the long stem sequences may participate in nonspecific target binding, indicating that these quadruplex MBs still require some improvements.

In addition to utilizing the regulatory effects of different concentrations of molecular stimuli, the design of MBs responsive to external physical triggers

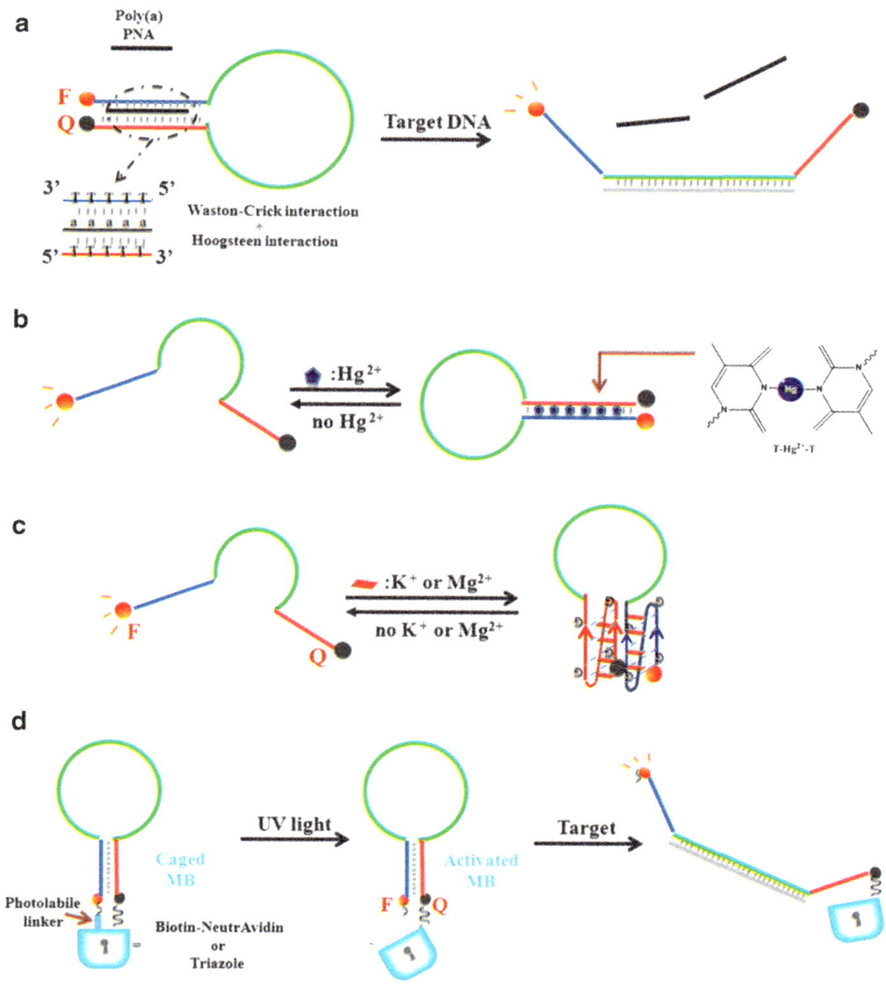

Fig. 7.5 Stem region engineering: (**a**) Triplex MBs based on T-a-T interactions; (**b**) Hg^{2+}—controlled MBs based on T-Hg^{2+}—T interactions; (**c**) cation-regulated G-quadruplex MBs; (**d**) caged MBs activated by light irradiation

is another promising direction for engineering. For example, functionalization of MBs with photoresponsive groups could provide exact temporal and spatial control over the functions of MBs. Compared with a chemically triggered response, photoregulation enables remote control of the timing, location, and intensity of the light "dosage," and it minimizes the adverse effects of other chemical components. The Yang group has realized this idea by developing "caged" MBs (Fig. 7.5d) [36]. In this design, biotin-avidin interaction or triazole linkages were utilized to lock the MBs' stems via a photocleavable linker (PC-linker) bearing an *o*-nitrobenzyl moiety. Opening of the "cage" by seconds of UV irradiation efficiently recovered the target binding affinity of the MBs. Besides photoinduced regulation, the design of other external triggers, even multiple triggers, can be a promising direction for future MB engineering.

7.4 Reporting Region Engineering

Generating highly sensitive MB probes can be a complex undertaking. Conventional MBs use organic fluorophores and quenchers, such as fluorescein or dabcyl. The beacon's sensitivity can be improved by either reducing the background signals in the closed hairpin state or increasing the fluorescence intensity of the open state. Incorporation of so-called superquenchers [37], which are actually assemblies of multiple quenching events, was one of the first ways to reduce background signals. After that, copper (Cu^{2+})-quenched [38], guanine base (G)-quenched [39], or nanoarchitecture (e.g., gold clusters [40], single-walled carbon nanotubes [41], and graphene oxide [42])-quenched MB structures were also developed. Similarly, the idea of enhancing fluorescence intensity of the open state has inspired a novel MB design using poly (phenylene ethynylene) (PPE) [43], quantum dots [44], and oligothiophene [45]. Readers interested in this research may refer other chapters of this book.

Some of the major drawbacks of conventional MBs based on fluorophore-quencher pair are residual fluorescence from incomplete quenching and limited sensitivity. To solve these problems, MBs with two fluorophores have been designed, instead of using a fluorophore-quencher pair. The fluorescence resonance energy transfer (FRET)-based MBs [46], excimer-monomer switching MBs [47], and self-quenched dimer-based MBs [48] are the examples of these quencher-free MB probes.

Other than multiple fluorophore labeling, quencher-free MBs were designed using a monolabeled beacon structure via changes in the microenvironment of the fluorophores [49]. Fluorescence quenching by nucleobases (normally G base) have been reported previously, mostly via photoinduced electron transfer between the nucleobases and the excited state of the fluorophore. However, other factors, such as the changes in solvent polarity, hydrophobicity around the fluorophores, and the pH of the media, must also be considered when designing quencher-free MBs. Taking advantage of well-developed fluorescent nucleotide synthesis methods, probes like HyBeacons [50], twin probes [51], and other base-discriminating fluorescent probes [52–54] have been generated. In the presence of target sequences, these probes display hybridization-induced fluorescence intensity changes. More importantly, single-nucleotide polymorphisms can be easily distinguished using these monolabeled probes. Typically, fluorescently labeled nucleobases are required for this application. Fluorophores, such as pyrene, anthracene, fluorescein, or Cy3, have been incorporated with deoxy-bases, U bases, or LNA for this application by incorporation into different regions of MB structures, predominantly the loop part [53].

In another interesting direction of study, these fluorescently labeled pseudo bases were incorporated into the stem region of the beacon probe, forming a so-called in-stem molecular beacon (ISMB) [55–58]. The fluorophore- and quencher-labeled bases in these ISMBs function as pseudo base pairs, which stack together in the middle of the stem domain in the closed beacon structure. As a result of the close stacking, the background fluorescence intensity is greatly suppressed. In addition, these interstrand dimers could be regarded as hydrophobically interacting base pairs to further stabilize the duplex stem regions. In a recent example, electron-rich pyrenes and electron-poor perylenediimides were synthesized into different stem arms of the MB probes [59]. The stacking of these pseudo base pairs results in very robust signal properties: high signal intensity, low background fluorescence, and red-shifted emission. DNA target as low as 0.3 nM could be detected using this beacon design.

Fluorescence detection is quite sensitive and functions as the typical signaling method for MB recognition events. However, applications can be hindered by the background signals from ubiquitous endogenous fluorescent components in the biological environment. Meanwhile, issues such as photobleaching and blinking, as well as toxicity, can also restrict applications. Nonfluorescence detection methods have thus begun to gain some attention for MB engineering. For example, phosphorescence with a long emission triplet lifetime could be suitable for reducing background signals, since an elongated delay time can avoid the influences of short-lived background fluorescence and scattering light. Li et al. developed a room-temperature phosphorescence-based MB [59], using a Eu^{3+} complex of chlorosulfonylated tetradentate ß-diketone and BHQ-2 quencher. More than 200-fold signal enhancement was achieved in the presence of nanomolar target sequence, as concentrations as low as 500 pM DNA could be directly detected in cell media.

Electrochemiluminescence (ECL), where the generation of optical signal is triggered by an electrochemical reaction, can also be employed. Wang et al. built a solid-state ECL film based on efficient ferrocene (Fc) quenching of ECL from ruthenium (II) tris-(bipyridine) ($Ru(bpy)_3{}^{2+}$) [60, 61]. Target hybridization with Fc-labeled MB results in a change in the ECL intensity, and the thermodynamic parameters of the MB binding and stem-loop formation can also be evaluated using this method.

By suitable modification, MBs can have functions beyond detection, most importantly in biomedical applications. As target-activatable probes, MBs offer controlled light emission in the presence of specific targets. The combination of photodynamic therapy (PDT) with MBs has triggered a new research concept for controlled phototherapy. PDT is a type of approved cell-killing process which relies on the generation of cytotoxic singlet oxygen ($^1O_2$) by light activation of a photosensitizer (PS). Zheng et al. introduced the concept of photodynamic MBs (PMB) to specifically control $^1O_2$ production to certain cancer biomarkers [15]. The PMB probe is a hairpin structure labeled at two ends by a PS and a quencher (Fig. 7.6a). The photoactivity of the PS is silenced until the linker interacts with a target molecule, such as matrix metalloproteinase-7 (MMP7) in their study. Thus, selective PDT-induced cell death could be achieved, depending on

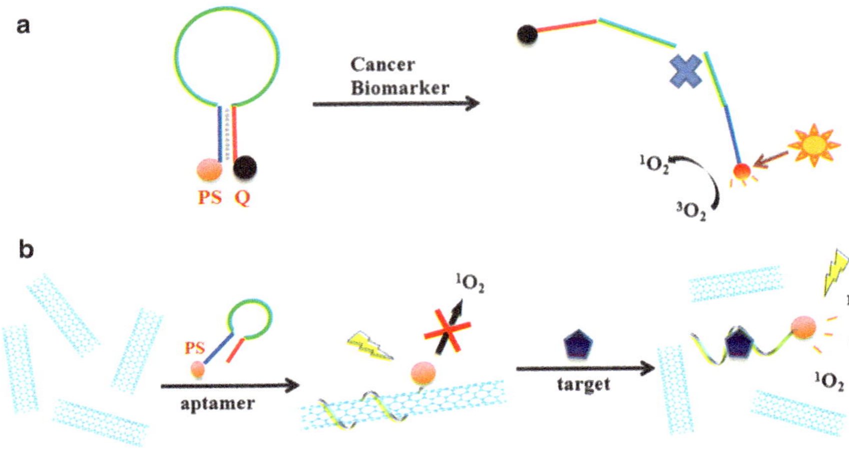

Fig. 7.6 Reporting region engineering: (**a**) Target-activated MBs for photodynamic therapy (PDT); (**b**) SWNT-based PDT-MB analogues

the presence of the biomarker (MMP7) in the target cancer cells. After this study, additional designs, including a dual-labeled self-quenching photosensitizer MB [62] and a multiconjugate "linear superquencher" [63], have been introduced to enhance signal-to-background ratio and reduce nontarget killing.

The Tan group has also developed a beacon-like structure, using SWNT quenching effects to regulate singlet oxygen generation (Fig. 7.6b) [64]. In the absence of the target molecule, PS-labeled DNA wraps onto the surface of the SWNT through – stacking, thereby quenching the PS fluorescence. In the presence of the target of interest, DNA-SWNT interaction is disrupted, causing DNA, together with PS, to leave the SWNT surface, followed by activation of phototherapy. More interestingly, the SWNT could also function as a cargo carrier for probe delivery to protect the DNA from enzymatic digestion or degradation in the biological environment. As these examples suggest, MBs can be engineered to perform multiple tasks in biological systems, not only diagnostics but therapeutic and other biological functions.

7.5 Conclusion and Future Perspectives

The past 16 years have witnessed the rapid development of a maturing MB technology. The highly efficient signal-transduction mechanism, versatile chemical modifications, and accessible and cost-effective synthesis have made MBs important components in diagnostic and therapeutic toolkits. This chapter has emphasized the engineering features currently used for equipping and refining MBs with novel and advanced functions to overcome theoretical and practical handicaps that inhibit their real applications.

In future years, the possibility of MBs for more nongene target detection, e.g., protein assays, enzyme monitoring, and whole cell detection, in biological systems will be investigated. Rational engineering pathways to boost the performance of MBs, equip them with adjunctive functions, and reduce the cost of large-scale MB modification will lead MB development into major research areas at the forefront of technology with significant commercial potential, including molecular computing [65], cancer cell identification, and disease diagnosis.

References

1. Bratu DP, Cha BJ, Mhlanga MM, Kramer FR, Tyagi S (2003) Visualizing the distribution and transport of mRNAs in living cells. Proc Natl Acad Sci U S A 100:13308–13313
2. Medley CD, Drake TJ, Tomasini JM, Rogers RJ, Tan W (2005) Simultaneous monitoring of the expression of multiple genes inside of single breast carcinoma cells. Anal Chem 77:4713–4718
3. Yang CJ, Wang L, Wu Y, Kim Y, Medley CD, Lin H, Tan W (2007) Synthesis and investigation of deoxyribonucleic acid/locked nucleic acid chimeric molecular beacons. Nucleic Acids Res 35:4030–4041
4. Tsourkas A, Behlke MA, Bao G (2002) Hybridization of 2′-O-methyl and 2′-deoxy molecular beacons to RNA and DNA targets. Nucleic Acids Res 30:5168–5174
5. Molenaar C, Marras SA, Slats JC, Truffert JC, Lemaître M, Raap AK, Dirks RW, Tanke HJ (2001) Linear 2′ O-Methyl RNA probes for the visualization of RNA in living cells. Nucleic Acids Res 29:E89–89
6. Wang L, Yang CJ, Medley CD, Benner SA, Tan W (2005) Locked nucleic acid molecular beacons. J Am Chem Soc 127:15664–15665
7. Crey-Desbiolles C, Ahn DR, Leumann CJ (2005) Molecular beacons with a homo-DNA stem: improving target selectivity. Nucleic Acids Res 33:e77
8. Browne KA (2005) Sequence-specific, self-reporting hairpin inversion probes. J Am Chem Soc 127:1989–1994
9. Kim Y, Yang CJ, Tan W (2007) Superior structure stability and selectivity of hairpin nucleic acid probes with an L-DNA stem. Nucleic Acids Res 35:7279–7287
10. Sheng P, Yang Z, Kim Y, Wu Y, Tan W, Benner SA (2008) Design of a novel molecular beacon: modification of the stem with artificially genetic alphabet. Chem Commun (Camb) 41:5128–5130
11. Zheng G, Chen J, Stefflova K, Jarvi M, Li H, Wilson BC (2007) Photodynamic molecular beacon as an activatable photosensitizer based on protease-controlled singlet oxygen quenching and activation. Proc Natl Acad Sci U S A 104:8989–8994
12. Yeh HY, Yates MV, Mulchandani A, Chen W (2008) Visualizing the dynamics of viral replication in living cells via Tat peptide delivery of nuclease-resistant molecular beacons. Proc Natl Acad Sci U S A 105:17522–17525
13. Chen AK, Behlke MA, Tsourkas A (2009) Sub-cellular trafficking and functionality of 2′-O-methyl and 2′-O-methyl-phosphorothioate molecular beacons. Nucleic Acids Res 37:e149
14. Xi C, Balberg M, Boppart SA, Raskin L (2003) Use of DNA and peptide nucleic acid molecular beacons for detection and quantification of rRNA in solution and in whole cells. Appl Environ Microbiol 69:5673–5678
15. Kuhn H, Demidov VV, Gildea BD, Fiandaca MJ, Coull JC, Frank-Kamenetskii MD (2001) PNA beacons for duplex DNA. Antisense Nucleic Acid Drug Dev 11:265–270
16. Seitz O (2000) Solid-phase synthesis of doubly labeled peptide nucleic acids as probes for the real-time detection of hybridization this work was supported by the Fonds der Chemischen Industrie. Angew Chem Int Ed Engl 39:3249–3252

17. Nitin N, Rhee WJ, Bao G (2009) Translation inhibition reveals interaction of 2′-deoxy and 2′-O-methyl molecular beacons with mRNA targets in living cells. Nucleic Acids Res 37:4977–4986
18. Mhlanga MM, Vargas DY, Fung CW, Kramer FR, Tyagi S (2005) tRNA-linked molecular beacons for imaging mRNAs in the cytoplasm of living cells. Nucleic Acids Res 33:1902–1912
19. Chen AK, Behlke MA, Tsourkas A (2007) Avoiding false-positive signals with nuclease-vulnerable molecular beacons in single living cells. Nucleic Acids Res 35:e105
20. Braasch DA, Corey DR (2001) Locked nucleic acid (LNA): fine-tuning the recognition of DNA and RNA. Chem Biol 8:1–7
21. Vester B, Wengel J (2004) LNA (locked nucleic acid): high-affinity targeting of complementary RNA and DNA. Biochemistry 43:13233–13241
22. Wu Y, Yang CJ, Moroz LL, Tan W (2008) Nucleic acid beacons for long-term real-time intracellular monitoring. Anal Chem 80:3025–3028
23. Egholm M, Nielsen PE, Buchardt O, Berg RH (1992) Recognition of guanine and adenine in DNA by cytosine and thymine containing peptide nucleic-acids (Pna). J Am Chem Soc 114:9677–9678
24. Gildea BD, Casey S, MacNeill J, Perry-O'Keefe H, Sorensen D, Coull JM (1998) PNA solubility enhancers. Tetrahedron Lett 39:7255–7258
25. Oh KJ, Cash KJ, Hugenberg V, Plaxco KW (2007) Peptide beacons: a new design for polypeptide-based optical biosensors. Bioconjug Chem 18:607–609
26. Oh KJ, Cash KJ, Plaxco KW (2006) Excimer-based peptide beacons: a convenient experimental approach for monitoring polypeptide-protein and polypeptide-oligonucleotide interactions. J Am Chem Soc 128:14018–14019
27. Socher E, Bethge L, Knoll A, Jungnick N, Herrmann A, Seitz O (2008) Low-noise stemless PNA beacons for sensitive DNA and RNA detection. Angew Chem Int Ed Engl 47:9555–9559
28. Thurley S, Röglin L, Seitz O (2007) Hairpin peptide beacon: dual-labeled PNA-peptide-hybrids for protein detection. J Am Chem Soc 129:12693–12695
29. Grossmann TN, Röglin L, Seitz O (2007) Triplex molecular beacons as modular probes for DNA detection. Angew Chem Int Ed Engl 46:5223–5225
30. Tang Z, Mallikaratchy P, Yang R, Kim Y, Zhu Z, Wang H, Tan W (2008) Aptamer switch probe based on intramolecular displacement. J Am Chem Soc 130:11268–11269
31. Frank-Kamenetskii MD, Mirkin SM (1995) Triplex DNA structures. Annu Rev Biochem 64:65–95
32. Yang R, Jin J, Long L, Wang Y, Wang H, Tan W (2009) Reversible molecular switching of molecular beacon: controlling DNA hybridization kinetics and thermodynamics using mercury(ii) ions. Chem Commun (Camb) 3:322–324
33. Lin YW, Ho HT, Huang CC, Chang HT (2008) Fluorescence detection of single nucleotide polymorphisms using a universal molecular beacon. Nucleic Acids Res 36:e123
34. Bourdoncle A, Estévez Torres A, Gosse C, Lacroix L, Vekhoff P, Le Saux T, Jullien L, Mergny JL (2006) Quadruplex-based molecular beacons as tunable DNA probes. J Am Chem Soc 128:11094–11105
35. Fu R, Li T, Lee SS, Park HG (2011) DNAzyme molecular beacon probes for target-induced signal-amplifying colorimetric detection of nucleic acids. Anal Chem 83:494–500
36. Wang C, Zhu Z, Song Y, Lin H, Yang CJ, Tan W (2011) Caged molecular beacons: controlling nucleic acid hybridization with light. Chem Commun (Camb) 47:5708–5710
37. Yang CJ, Lin H, Tan W (2005) Molecular assembly of superquenchers in signaling molecular interactions. J Am Chem Soc 127:12772–12773
38. Brunner J, Kraemer R (2004) Copper(II)-quenched oligonucleotide probes for fluorescent DNA sensing. J Am Chem Soc 126:13626–13627
39. Saito Y, Mizuno E, Bag SS, Suzuka I, Saito I (2007) Design of a novel G-quenched molecular beacon: a simple and efficient strategy for DNA sequence analysis. Chem Commun (Camb) 43:4492–4494
40. Dubertret B, Calame M, Libchaber AJ (2001) Single-mismatch detection using gold-quenched fluorescent oligonucleotides. Nat Biotechnol 19:365–370

41. Yang R, Jin J, Chen Y, Shao N, Kang H, Xiao Z, Tang Z, Wu Y, Zhu Z, Tan W (2008) Carbon nanotube-quenched fluorescent oligonucleotides: probes that fluoresce upon hybridization. J Am Chem Soc 130:8351–8358
42. Lu CH, Li J, Liu JJ, Yang HH, Chen X, Chen GN (2010) Increasing the sensitivity and single-base mismatch selectivity of the molecular beacon using graphene oxide as the "nanoquencher". Chemistry 16:4889–4894
43. Yang CJ, Pinto M, Schanze K, Tan W (2005) Direct synthesis of an oligonucleotide-poly(phenylene ethynylene) conjugate with a precise one-to-one molecular ratio. Angew Chem Int Ed Engl 44:2572–2576
44. Yeh HY, Yates MV, Mulchandani A, Chen W (2010) Molecular beacon-quantum dot-Au nanoparticle hybrid nanoprobes for visualizing virus replication in living cells. Chem Commun (Camb) 46:3914–3916
45. Cazzato A, Capobianco ML, Zambianchi M, Favaretto L, Bettini C, Barbarella G (2007) Oligothiophene molecular beacons. Bioconjug Chem 18:318–322
46. Zhang P, Beck T, Tan W (2001) Design of a molecular beacon DNA probe with two fluorophores. Angew Chem Int Ed Engl 40:402–405
47. Conlon P, Yang CJ, Wu Y, Chen Y, Martinez K, Kim Y, Stevens N, Marti AA, Jockusch S, Turro NJ, Tan W (2008) Pyrene excimer signaling molecular beacons for probing nucleic acids. J Am Chem Soc 130:336–342
48. Nesterova IV, Erdem SS, Pakhomov S, Hammer RP, Soper SA (2009) Phthalocyanine dimerization-based molecular beacons using near-IR fluorescence. J Am Chem Soc 131:2432–2433
49. Venkatesan N, Seo YJ, Kim BH (2008) Quencher-free molecular beacons: a new strategy in fluorescence based nucleic acid analysis. Chem Soc Rev 37:648–663
50. Dobson N, McDowell DG, French DJ, Brown LJ, Mellor JM, Brown T (2003) Synthesis of HyBeacons and dual-labelled probes containing 2′-fluorescent groups for use in genetic analysis. Chem Commun (Camb) 11:1234–1235
51. Ergen E, Weber M, Jacob J, Herrmann A, Müllen K (2006) Twin probes as a novel tool for the detection of single-nucleotide polymorphisms. Chemistry 12:3707–3713
52. Østergaard ME, Cheguru P, Papasani MR, Hill RA, Hrdlicka PJ (2010) Glowing locked nucleic acids: brightly fluorescent probes for detection of nucleic acids in cells. J Am Chem Soc 132:14221–14228
53. Martí AA, Jockusch S, Li Z, Ju J, Turro NJ (2006) Molecular beacons with intrinsically fluorescent nucleotides. Nucleic Acids Res 34:e50
54. Hwang GT, Seo YJ, Kim BH (2004) A highly discriminating quencher-free molecular beacon for probing DNA. J Am Chem Soc 126:6528–6529
55. Häner R, Biner SM, Langenegger SM, Meng T, Malinovskii VL (2010) A highly sensitive, excimer-controlled molecular beacon. Angew Chem Int Ed Engl 49:1227–1230
56. Hara Y, Fujii T, Kashida H, Sekiguchi K, Liang XG, Niwa K, Takase T, Yoshida Y, Asanuma H (2010) Coherent quenching of a fluorophore for the design of a highly sensitive in-stem molecular beacon. Angew Chem Int Ed 49:5502–5506
57. Berndl S, Wagenknecht HA (2009) Fluorescent color readout of DNA hybridization with thiazole orange as an artificial DNA base. Angew Chem Int Ed Engl 48:2418–2421
58. Kashida H, Takatsu T, Fujii T, Sekiguchi K, Liang X, Niwa K, Takase T, Yoshida Y, Asanuma H (2009) In-stem molecular beacon containing a pseudo base pair of threoninol nucleotides for the removal of background emission. Angew Chem Int Ed Engl 48:7044–7047
59. Li J, Zhou W, Ouyang X, Yu H, Yang R, Tan W, Yuan J (2011) Design of a room-temperature phosphorescence-based molecular beacon for highly sensitive detection of nucleic acids in biological fluids. Anal Chem 83:1356–1362
60. Wang X, Yun W, Dong P, Zhou J, He P, Fang Y (2008) A controllable solid-state Ru(bpy)(3)(2+) electrochemiluminescence film based on conformation change of ferrocene-labeled DNA molecular beacon. Langmuir 24:2200–2205

61. Wilson R, Johansson MK (2003) Photoluminescence and electrochemiluminescence of a Ru(II)(bpy)3-quencher dual-labeled oligonucleotide probe. Chem Commun (Camb) 3:2710–2711
62. Gao Y, Qiao G, Zhuo L, Li N, Liu Y, Tang B (2011) A tumor mRNA-mediated bi-photosensitizer molecular beacon as an efficient imaging and photosensitizing agent. Chem Commun (Camb) 47:5316–5318
63. Lovell JF, Chen J, Huynh E, Jarvi MT, Wilson BC, Zheng G (2010) Facile synthesis of advanced photodynamic molecular beacon architectures. Bioconjug Chem 21:1023–1025
64. Zhu Z, Tang Z, Phillips JA, Yang R, Wang H, Tan W (2008) Regulation of singlet oxygen generation using single-walled carbon nanotubes. J Am Chem Soc 130:10856–10857
65. Seelig G, Soloveichik D, Zhang DY, Winfree E (2006) Enzyme-free nucleic acid logic circuits. Science 314:1585–1588

Chapter 8
Excimer Molecular Beacon

Jin Huang

Abstract During the detection in complex biological system, similar to other fluorescence probes, the molecular beacons suffer severely from the background signal interference. Recent studies indicated that excimer molecular beacon (EMB) can address the problem. EMB is a dual-pyrene-labeled hairpin DNA structure with large Stokes shift and long fluorescence lifetime, which afford an effective strategy for detection in complex biological environment. In the chapter, the recent development of the research of EMB is presented. Firstly, the general design of EMB, as well as the structure and working mechanism, is introduced. Furthermore, the synthesis and properties of EMB are descripted, which explain its capability of detection in complex environment and high sensitivity and selectivity. Finally, the examples of the different applications are discussed, including nucleic acids and other molecules detection.

8.1 Introduction

When applied to its native environment, a molecular beacon confronts two significant background signal sources. The first is the probe itself. For example, when a quenching-based MB is used for analysis, the probe always has some incomplete quenching, resulting in a significant probe background. The second comes from the native fluorescence of the biological environment. There are many molecular species in a biological environment, some of which will give a strong fluorescence background signal upon excitation. These problems deteriorate assay sensitivity, compromise probe selectivity, and thus hinder the analysis of small

J. Huang (✉)
State Key Laboratory of Chemo/Biosensing and Chemometrics, College of Chemistry and Chemical Engineering, Hunan University, Changsha 410082, China
e-mail: jinhuang@hnu.edu.cn

C.J. Yang and W. Tan (eds.), *Molecular Beacons*, DOI 10.1007/978-3-642-39109-5_8,

molecules, nucleic acids, and proteins in a complex sample. Thus, two challenges are raised: (a) how to minimize background signal and (b) how to tolerate complex biological sample interference.

To address these issues, FRET pairs [1–5], metal nanoparticles [6–8], and pyrene moieties [9–11] have been incorporated into MBs and have, to a certain degree, improved the properties of traditional MBs for detection in a complex environment. Among these, pyrene-based excimer-forming MBs (Py-MBs), which possess the advantage of long fluorescence lifetime, can discriminate the long-lived pyrene fluorescence signal from the short-lived fluorescence background. It therefore seems to be a promising strategy for detecting analytes in complex biological environment. This chapter will discuss the excimer molecular beacon (EMB), including its design, synthesis, properties, and applications.

8.2 Design of Excimer Molecular Beacon

A novel EMB, first designed by Fujimoto et al. in 2004, was developed on the basis of a monomer–excimer switch [9]. Instead of a fluorophore-quencher pair, pyrene monomers were labeled at opposite termini of the MB to facilitate an excimer–monomer switching mechanism that is controlled by bound and unbound conformations of the oligonucleotide. Pyrene is a polycyclic aromatic hydrocarbon (PAH) consisting of four fused benzene rings, resulting in a flat aromatic system. Pyrene forms during incomplete combustion of organic compounds. Figure 8.1

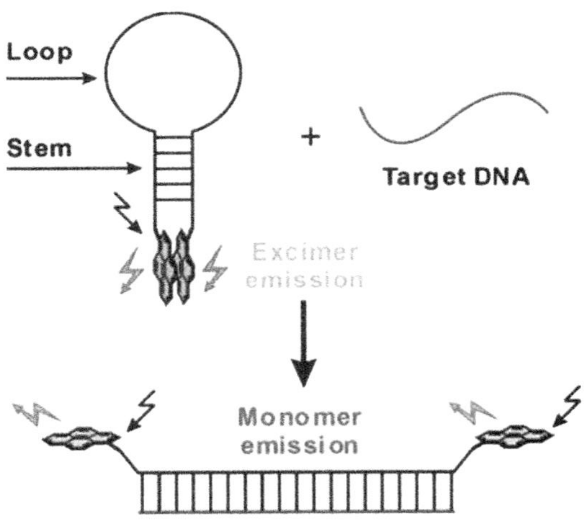

Fig. 8.1 Schematic representation of excimer molecular beacon (*EMB*), which was dually labeled with pyrene at the 3′- and 5′-end of single-stranded oligonucleotides with a stem-loop structure. In the absence of target DNAs, the stem-close-shaped EMB predominantly emits excimer fluorescence (*yellow-green*). However, upon hybridization with target DNAs, the EMB undergoes a dynamic conformational change to emit monomer fluorescence (*pale blue*) (Reprinted from Ref. [9]. Copyright 2008, with permission from Elsevier) (Color figure online)

shows the working principle of the typical EMB, where two pyrene derivatives are modified on the 5′- and 3′-end of one single oligonucleotide, respectively. In the absence of target DNA, the hairpin structure brings the two pyrene moieties into close proximity and allows the formation of an excimer that emits fluorescence. In the presence of target DNA, the hairpin structure undergoes conformational change upon hybridization, the two pyrene molecules are spatially separated, and only the monomer peaks can be observed. In this way, the emission wavelength shift can fully minimize the background signal, while, at the same time, the large difference in fluorescence lifetime between pyrene excimer and the background fluorescence lifetime makes it possible to apply this Py-MB in complex biological sample.

8.3 Synthesis of Excimer Molecular Beacon

EMB synthesis depends on modification of the pyrene moieties at the two termini of single-stranded oligonucleotides. According to the literature, the two methods of accomplishing this are (a) on-machine and (b) off-machine synthesis. As shown in Fig. 8.2, Fujimoto et al. synthesized a dual-labeled pyrene beacon, in which a single-stranded 29-mer DNA possessing a C_3-alkylamino linker (3′-end) was first coupled with a pyrene phosphoramidite (Fig. 8.2 (1)) at the 5′-end on a DNA synthesizer. After cleavage from the solid support and purification by HPLC, the 5′-modified probe was further tethered to pyrene succinimidyl ester (Fig. 8.2 (2)) at the amino linker of the 3′-end, thereby demonstrating a combination of on- and off-machine synthesis to generate this EMB, as discussed in detail below [9].

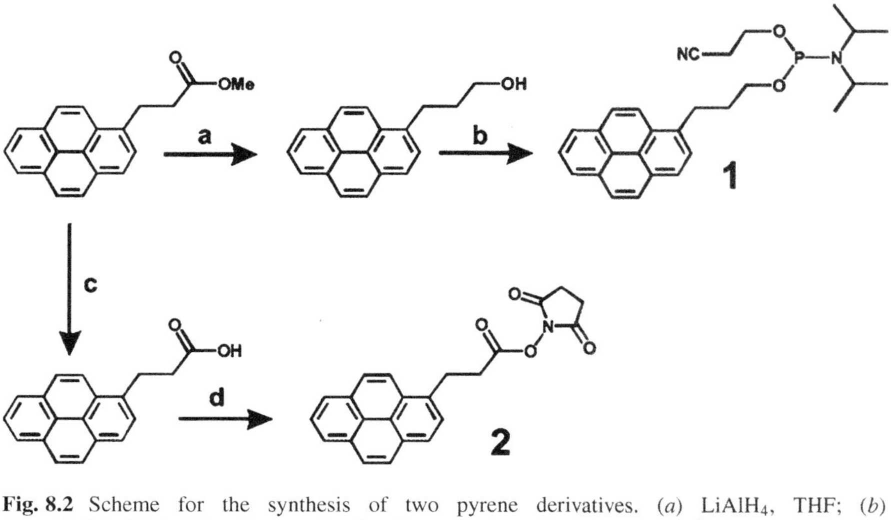

Fig. 8.2 Scheme for the synthesis of two pyrene derivatives. (*a*) $LiAlH_4$, THF; (*b*) $(i\text{-}Pr_2N)_2PO(CH_2)_2CN$, 1H-tetrazole, CH_3CN; (*c*) KON, CH_3OH; (*d*) *N,N*′-disuccinimidyl carbonate, pyridine, CH_3CN (Reprinted from Ref. [9]. Copyright 2008, with permission from Elsevier)

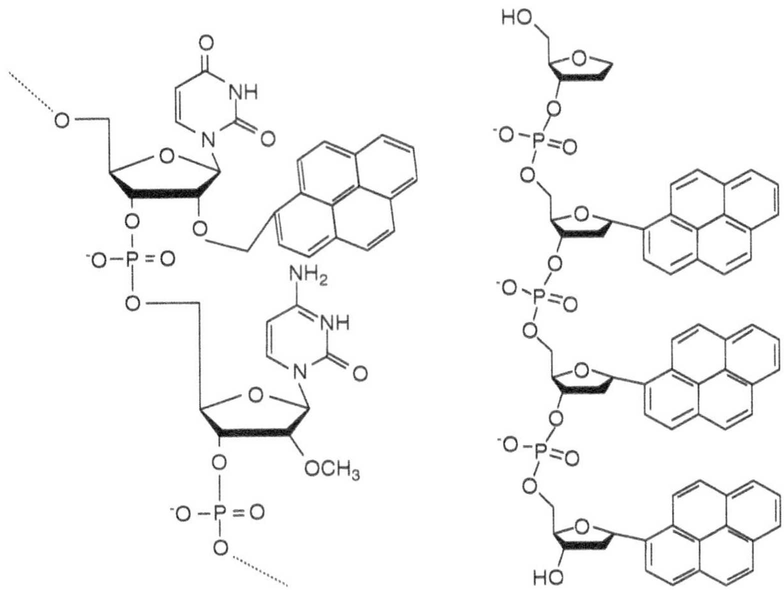

Fig. 8.3 Chemical structures of pyrene-conjugated 2′-O-methyloligoribonucleotide (*left*, reprinted from Ref. [12] by permission of Oxford University Press) and nonnatural tri-pyrene nucleoside analogue (*right*, reproduced from Ref. [13] by permission of The Royal Society of Chemistry)

8.3.1 On-Machine Synthesis

Oligonucleotide synthesis has been fully automated since the 1970s. Briefly, it involves the chemical synthesis of relatively short fragments of nucleic acids with defined chemical structure (sequence). The technique is extremely useful in current laboratory practice because it provides rapid and inexpensive access to custom-made oligonucleotides of the desired sequence. Chemical oligonucleotide synthesis is carried out in the 3′ to 5′ direction. Currently, the process is implemented as solid-phase synthesis using the phosphoramidite method and phosphoramidite building blocks derived from protected 2′-deoxynucleosides (dA, dC, dG, and T), or chemically modified nucleosides, e.g., LNA. To obtain the desired oligonucleotide, the building blocks are sequentially coupled to the growing oligonucleotide chain in the order required by the sequence of the product. In our case, the first step is to make a pyrene phosphoramidite building block that can be recognized by a DNA/RNA synthesizer. Yamana et al. introduced pyrene phosphoramidite to make oligonucleotide–pyrene conjugates such that a pyrene moiety was covalently linked to a nucleotide through a one-carbon spacer to the 2′-*O*-position of a ribose unit (Fig. 8.3, left) [12]. Kool et al. developed a class of nonnatural DNA nucleoside analogue, in which the DNA base is replaced at the C-1 position of ribose by the fluorescent pyrene dyes (Fig. 8.3, right) [13–15]. On-machine synthesis has high coupling efficiency but needs complex precursors, which therefore limits its use by researchers.

8.3.2 Off-Machine Synthesis

Off-machine synthesis has been widely used to conjugate pyrene molecules to oligonucleotide because of its simplicity. In these coupling reactions, either the 1-pyrene acetic acid or succinimidyl ester of 1-pyrene butanoic acid is frequently used as the labeling reagents. Turro et al. labeled a pyrene-2′-methoxy RNA probe through the solution-phase synthesis method [16]. Huang et al. successfully synthesized EMBs by the same protocol. Here, amine-modified DNA probe was synthesized using standard phosphoramidite chemistry, and probes were purified using reversed-phase HPLC. The purified DNAs were conjugated with the pyrene at the amino end by incubation in 0.1 M sodium bicarbonate/sodium carbonate buffer (pH 8.5) mixed with a 15-fold excess of the succinimidyl ester of 1-pyrene butanoic acid dissolved in dimethyl sulfoxide (DMSO). The unconjugated pyrene was removed by ethanol precipitation of DNA. The resulting crude mixture was purified by HPLC again [17, 18].

Different from the above solution-phase synthesis method, Yang et al. developed a solid-phase synthesis method to dually label DNA sequences with pyrene [19]. The synthesis started with 3′ amine CPG and ended with an alkylamine linker. Two amine groups at the 3′- and 5′-ends were deprotected with piperidines and TCA, respectively. After this step, DNA bases were still protected and linked on solid support, while two amines at either end of the sequence were activated and ready to react with carboxylated pyrene. After the coupling reaction, the probe was treated and processed as a regular DNA sequence in deprotection, desalting, and purification procedures. In this way, only one purification step is needed for the whole synthesis, and coupling efficiency is much higher on solid support compared to that in solution.

8.4 Properties of Excimer Molecular Beacon

Some properties of EMBs are similar to MBs, such as thermodynamics, kinetics, and single-base mismatch discrimination. However, other properties are unique by the presence of pyrene. Pyrene is a polycyclic aromatic hydrocarbon consisting of four fused benzene rings, resulting in a flat aromatic system. Its electronic spectrum and state assignments have been well established for years. Pyrene has several special photophysical characteristics which make it extraordinarily suitable as a fluorescent probe. These include a high quantum yield of fluorescence ($\varphi = 0.6$) and a relatively long S1 lifetime ($\tau = 450$ ns in deaerated cyclohexane at room

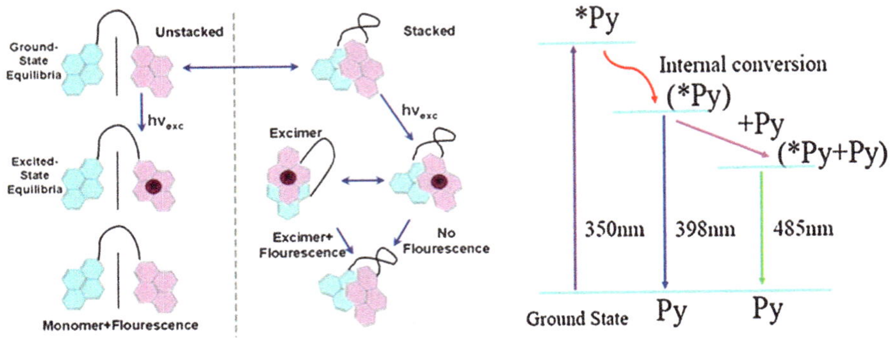

Fig. 8.4 Schematic illustration of the production of intramolecular excited dimer fluorescence from excited monomer pyrene: structural diagram (*left*) and simplified Jablonski-like diagram (*right*)

temperature), both readily measurable using straightforward techniques. Moreover, the relative intensities of the vibrational structure of pyrene's fluorescence spectrum are sensitive to the polarity of the surrounding environment.

In addition to emitting fluorescence from the excited monomer state, pyrene and its derivatives can also form an excited-state dimer or excimer. In our case, an excimer is formed if an excited pyrene monomer, during its fluorescence lifetime, interacts in a specific manner with a neighboring ground-state (unexcited) pyrene (Fig. 8.4). Excimer formation also depends on the environment in which the neighboring two pyrene molecules interact, for example, the solvent and the distances between the molecules. The pyrene molecules labeled on biomolecules should be in close proximity and should be able to rotate to achieve a precisely symmetrical configuration. Pyrene has a broad, featureless excimer emission centered at 480–500 nm, while the monomer emits at 380 and 400 nm. The large wavelength separation between excimer excitation (340 ~ 350 nm) and emission facilitates separating the true signal from background noise, while employing pyrene in a probe [20].

8.5 Applications of Excimer Molecular Beacon

8.5.1 Excimer Molecular Beacon for Nucleic Acids

The pyrene excimer-based molecular beacon has attracted much attention in the development of nucleic acid detection. Yamana et al. designed a novel bis-pyrene-labeled oligonucleotide probe (BP probe) for the detection of a single-base mismatch in single-strand DNA as a target (Fig. 8.5). The probe sequence was chosen to form a stem-loop structure similar to a molecular beacon probe (MB probe) and thus termed as BP–MB probe. Hybridization of the BP–MB probe with

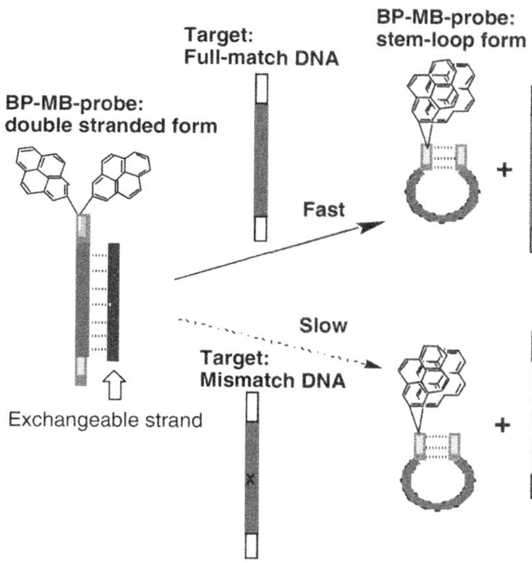

Fig. 8.5 Detection of DNA-base alteration using bis-pyrene-labeled oligonucleotide probe (Reprinted from Ref. [10]. Copyright 2008, with permission from Elsevier)

oligonucleotides provided the partially double-stranded BP–MB probe, in which the oligonucleotide sequences are complementary to the loop segment, but not to the stem sequence, and displaceable by target DNA. It was shown that these partial double-stranded BP–MB probes in the stem-loop structure display strong excimer fluorescence, but upon target recognition, they exhibit monomer fluorescence. The strand exchange reactions in the presence of cationic comb-type copolymer as a catalyst were monitored by the excimer fluorescence changes. The existence of a mismatched base can be determined by the slower rates compared with fully matched DNA [10].

Conlon et al. engineered molecular beacon DNA probes, containing 1–4 pyrene monomers on the 5′-end and quencher DABCYL on the 3′-end, for real-time probing of DNA sequences (Fig. 8.6). In the absence of a target sequence, the multiple-pyrene-labeled molecular beacons assumed a stem-closed conformation resulting in quenching of the pyrene excimer fluorescence. In the presence of target, the beacons switched to a stem-open conformation, which separated the pyrene label from the quencher molecule and generated an excimer emission signal proportional to the target concentration. Steady-state fluorescence assays resulted in a subnanomolar limit of detection in buffer, whereas time-resolved signaling enabled low-nanomolar target detection in cell-growth media. It was found that the excimer emission intensity could be scaled by increasing the number of pyrene monomers conjugated to the 5′-terminus. Each additional pyrene monomer resulted in substantial increases in the excimer emission intensities, quantum yields, and excited-state lifetimes of the hybridized MBs. The long fluorescence lifetime ($\sim$40 ns), large Stokes shift (130 nm), and tunable intensity of the excimer make this multiple-pyrene moiety a useful alternative to tradition fluorophore labeling in nucleic acid probes [11].

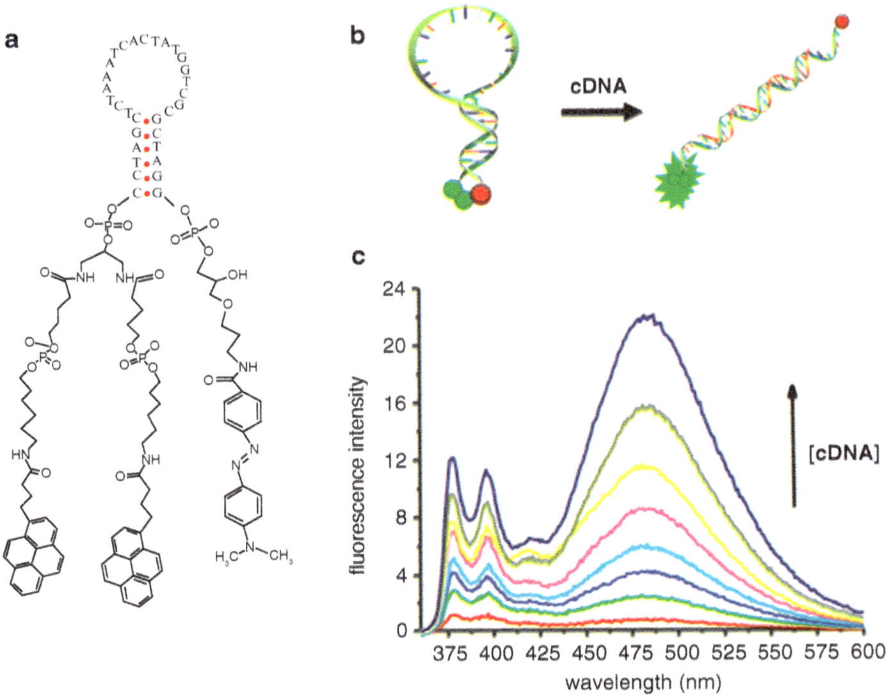

Fig. 8.6 (**a**) Structural representation of two pyrene molecules labeled to dual 5′ linker chains and DABCYL molecule labeled to 3′ chain (MB-2P). (**b**) Schematic of MB-2P hybridization with complementary target. (**c**) Emission spectra of 1-μM MB-2P with increasing concentrations of cDNA in buffer solution (Reprinted with the permission from Ref. [11]. Copyright 2008 American Chemical Society)

Häner et al. proposed a molecular beacon in which signal control is accomplished by formation of a donor–acceptor (D–A) complex (Fig. 8.7). The stem contains pairs of nonnucleic pyrenes (Y) and perylene diimides (PDIs, E) that can interact by interstrand stacking. In the native structure, this leads to efficient signal suppression, whereas the hybridized form is characterized by an excimer signal produced by the two adjacent pyrenes. Additionally, the formation of a stable D–A complex helps to minimize the number of natural bases in the stem, thus reducing the changes of unwanted base-pairing interactions [21].

Huang et al. described an amplified DNA-detection system which employed a two-pyrene dual-labeled hairpin structure (H1* and H2*, Fig. 8.8). Each hairpin has a stem of 18 base pairs enclosing a 6 nucleotide (nt) loop. Each also has an additional 6 nt sticky end at the 5′-end of H1* (complementary to the loop of H2*) and at the 3′-end of H2* (complementary to the loop of H1*). In the absence of target DNA, both probes, H1* and H2*, are in the closed form, and the two pyrene moieties are spatially separated by the extra length of the sticky end. In this state, only the

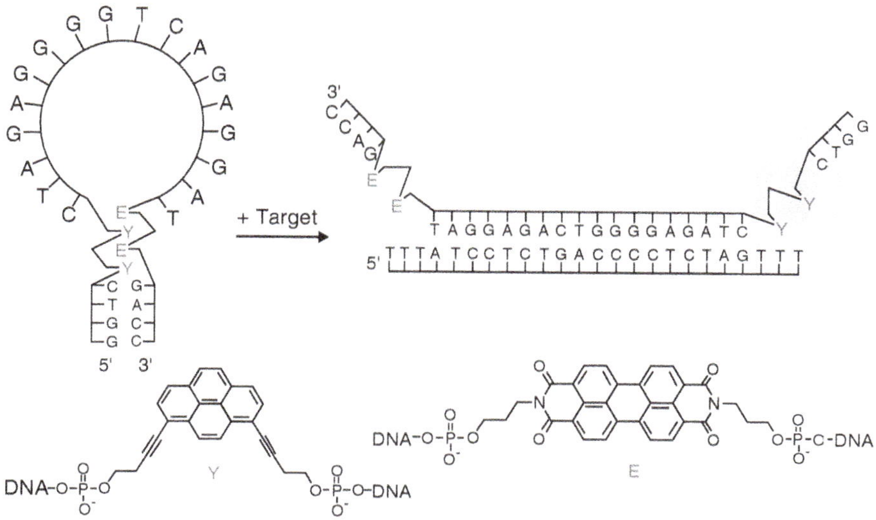

Fig. 8.7 Illustration of an excimer-controlled molecular beacon; excimer formation between pyrene derivatives (Y) is prevented by donor–acceptor complex formation with perylene diimides (PDI, E) in the native form. Upon hybridization with the target, PDI and pyrene units are separated, enabling excimer formation (Reproduced from Ref. [21] by permission of John Wiley & Sons Ltd.)

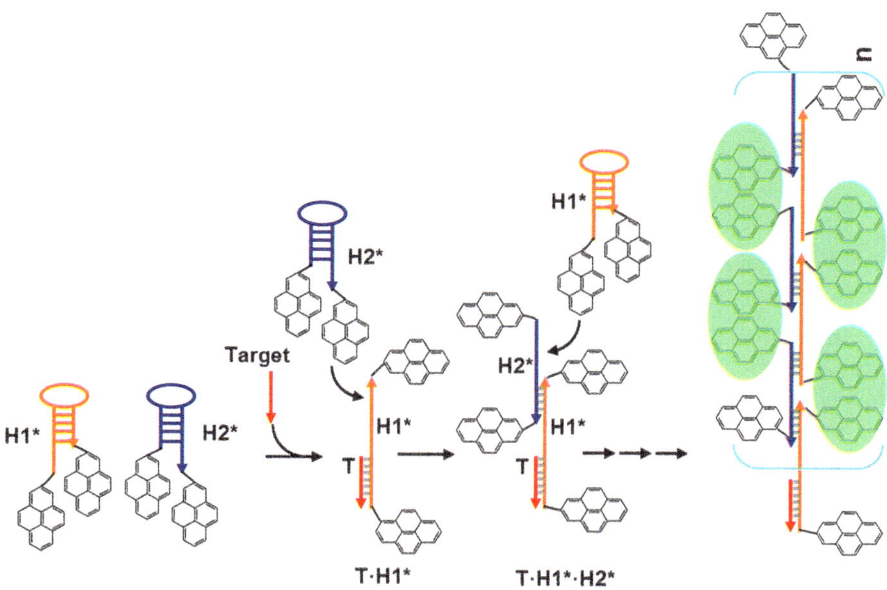

Fig. 8.8 Working principle behind the detection of DNA on the basis of HCR amplification and the formation of pyrene excimers (Reproduced from Ref. [18] by permission of John Wiley & Sons Ltd.)

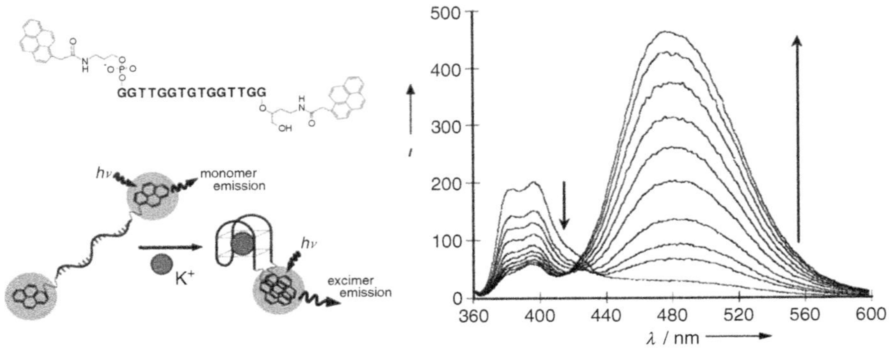

Fig. 8.9 The chemical structure, expected excimer formation (*left*), and fluorescence titration spectra of PSO-py (0.2 μM) with KCl (0–200 nM) in Tris–HCl solution (5 mM, pH 7.2) at 25 °C (Reproduced from Ref. [22] by permission of John Wiley & Sons Ltd.)

monomer emission peaks (at 375 and 398 nm) are observed. However, when the target is present in the solution, it pairs with the sticky end of H1*, which undergoes an unbiased strand-displacement interaction to open the hairpin. The newly exposed sticky end of H1* nucleates at the sticky end of H2* and opens the hairpin to expose a sticky end on H2*. This sticky end is identical in sequence to the target. In this way, each copy of the target can propagate a chain reaction of hybridization events between alternating H1* and H2* hairpins to form a nicked double helix. In this state, a pyrene moiety on one probe is brought into close proximity to a pyrene moiety on the neighboring probe. Thus, numerous pyrene excimers are formed, each emitting at approximately 485 nm. By observation of the emission intensities of the pyrene monomer and the excimer, the target DNA can be detected with high sensitivity [18].

8.5.2 Excimer Molecular Beacon for Other Molecules

Apart from nucleic acids, the pyrene excimer-based molecular beacon has been a promising probe for many important biological elements, ranging from small ions to biomacromolecules. Nagatoishi et al. reported a pyrene-labeled potassium-sensing oligonucleotide (PSO-py, Fig. 8.9), in which two pyrene molecules are attached to the termini of a 15-mer oligonucleotide (thrombin-binding aptamer, TBA), which can form stable G-quadruplex complexes upon binding to thrombin protein in the presence of K^+ ions with a 1:1 stoichiometry. Therefore, the binding of K^+ ions should organize the 5′ and 3′ termini of TBA so that the two attached pyrene moieties are arranged face to face. Thus, the pyrene moieties are expected to produce efficient excimer emission [22].

Yang et al. engineered light-switching pyrene excimer aptamers for rapid protein monitoring in complex biological fluids. The aptamer sequence that binds with high affinity to the target protein platelet-derived growth factor (PDGF)–BB is labeled

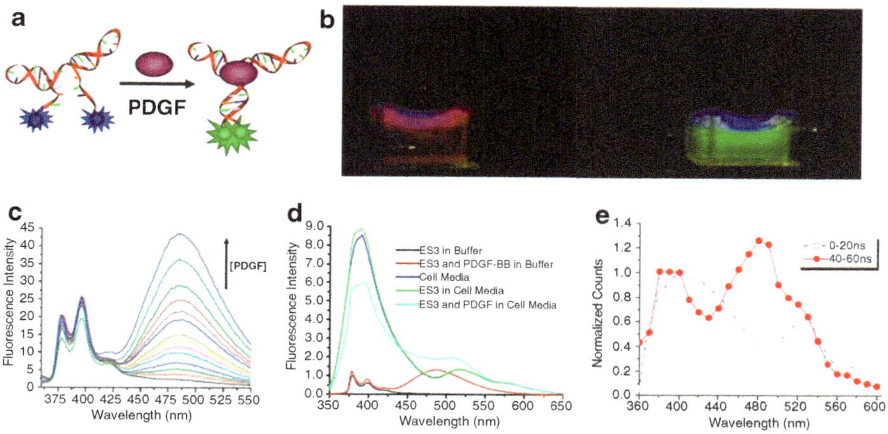

Fig. 8.10 Detection of PDGF in biological fluids. (**a**) Use of the pyrene excimer to probe PDGF. After binding to PDGF, pyrene excimer forms and changes the color from *blue* to *green*. (**b**) Visual detection of 4 pmol of PDGF–BB after illumination with a UV lamp. Solution of the 100 nM excimer probe without (*left*) and with (*right*) 40 nM of PDGF–BB. (**c**) Response of the excimer probe to different concentrations of PDGF–BB (0–40nM). (**d**) Steady-state fluorescence spectra and (**e**) time-resolved fluorescence spectra of PDGF in dyed cell media (Reprinted from Ref. [19]. Copyright 2005, National Academy of Sciences, U.S.A.)

with pyrene molecules at both ends. The specific binding of aptamer to target protein changes the conformation of the aptamer probe, bringing the two pyrene molecules into close proximity to form an excimer, which results in a change of fluorescence wavelength from 400 nm for the pyrene monomer to 485 nm for the pyrene excimer (Fig. 8.10). Combining light-switching and time-resolved measurements, picomolar PDGF–BB can be detected in a few seconds. Direct detection and quantification of target molecules in complex biological samples, such as a cultured cell dish, can be carried out without any need for sample cleanup [19].

Huang et al. engineered a competition-mediated pyrene-switching aptasensor for lysozyme detection in buffer or human serum using both steady-state and time-resolved measurement (Fig. 8.11). The approach involves two DNA strands, one a lysozyme aptamer and the other a dual-pyrene-labeled hairpin sequence (competitor), which is partially complementary to the aptamer. In the absence of target lysozyme, the aptamer hybridizes with part of competitor. As a result, the DNA hairpin opens, and both pyrene molecules are spatially separated. However, in the presence of the target, the competitor is displaced from the aptamer by the target, subsequently forming an initial hairpin structure. This brings the two pyrene moieties into close proximity to form an excimer, which, in turn, results in a significant shift of fluorescence emission from monomer to excimer. Meanwhile, with time-resolved emission measurements, the pyrene excimer signal can be separated from biological background interference [17].

Zheng et al. proposed a way to restrict the labeled pyrene molecules in a hydrophobic cavity of cyclodextrin (Fig. 8.12). This bonding, which acts like extra

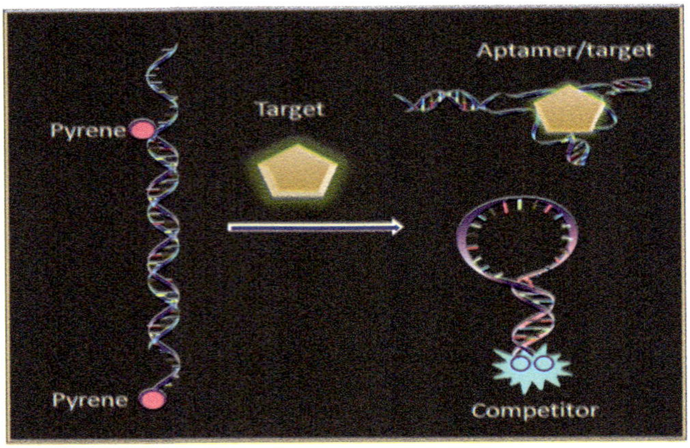

Fig. 8.11 Working principle of competition-mediated pyrene-switching aptasensor for an aptamer/target binding assay (Reprinted with the permission from Ref. [17]. Copyright 2010 American Chemical Society)

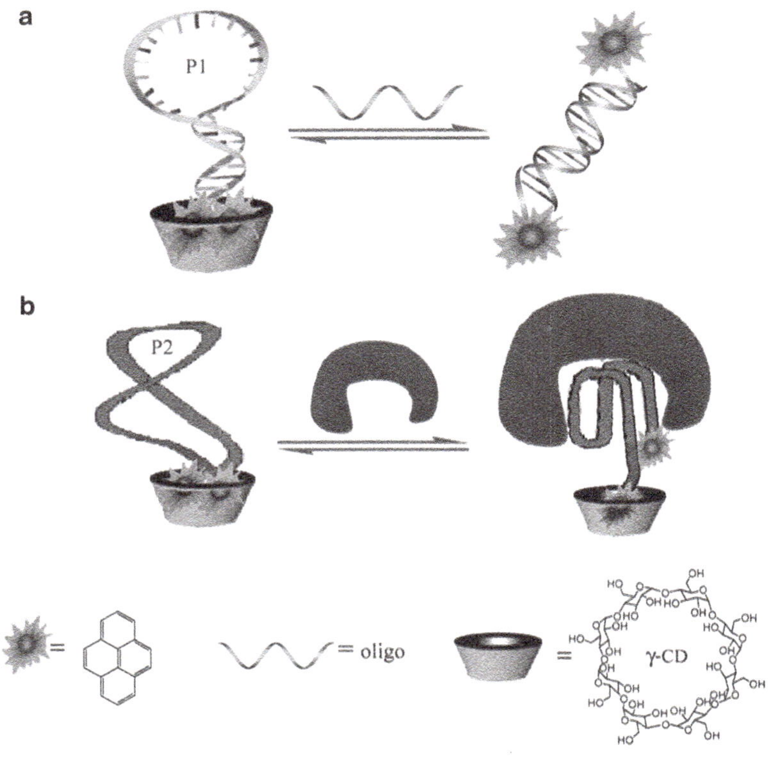

Fig. 8.12 Schematic representation of the cyclodextrin-bonded excimer–monomer switching probes for the detection of target DNA (**a**) and protein (**b**) (Reprinted with the permission from Ref. [23]. Copyright 2010 American Chemical Society)

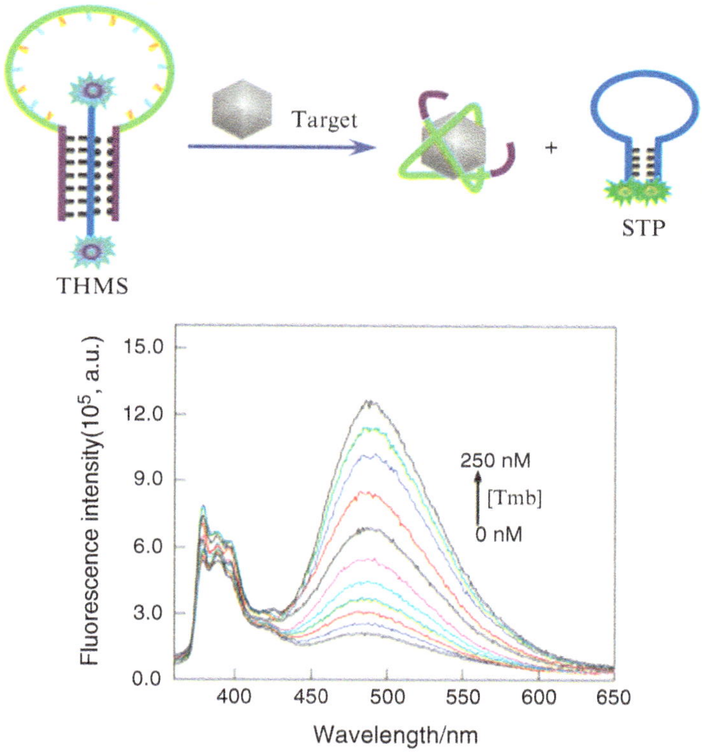

Fig. 8.13 *Top*: Scheme of triple-helix molecular switch (*THMS*) for signaling aptamer/target binding event. *Bottom*: Fluorescence emission spectra of THMS in the presence of increasing amounts of thrombin (Reprinted with the permission from Ref. [24]. Copyright 2010 American Chemical Society)

base pairs to form Watson–Crick duplex, achieves variation in the level of space proximity of the two labels and, thus, the degree of conformational constraint. To demonstrate the feasibility of the design, a stem-containing oligonucleotide probe (P1) for DNA hybridization assay and a stemless one (P2) for protein detection were examined as models. Both oligonucleotides were doubly labeled with pyrene at the 5′- and 3′-ends, respectively. It is the cyclodextrin/pyrene inclusion interaction that allows modulating the degree of conformation constraints of P1 and P2 and thus their background signals and selectivity [23]. After that, Zheng et al. reported a triple-helix molecular switch (THMS) based on Watson–Crick and Hoogsteen base pairings (Fig. 8.13). The THMS consists of a central, target-specific aptamer sequence flanked by two arm segments and a dual-labeled oligonucleotide serving as a signal transduction probe (STP). Aptamer/target binding results in the formation of a structured aptamer/target complex which disassembles the THMS and the STP [24].

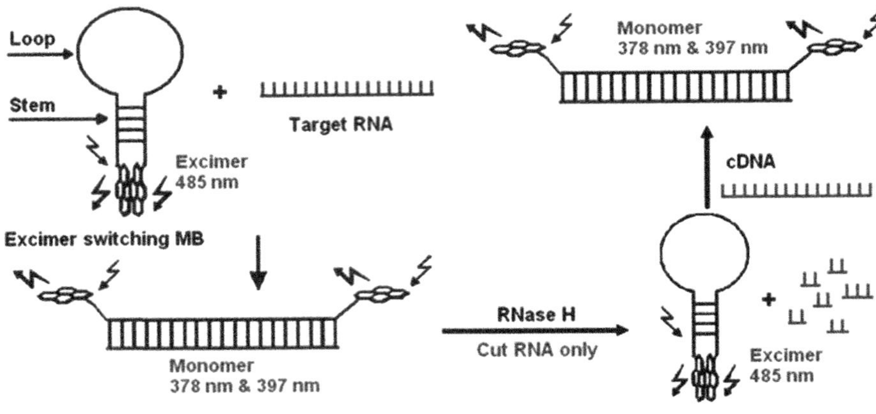

Fig. 8.14 Schematic representation of a light-switching excimer beacon probe to monitor RNase H activity (Reproduced from Ref. [25] by permission of John Wiley & Sons Ltd.)

With unique properties, dual-pyrene-labeled MBs also find interesting applications in enzyme activity studies. Chen et al. described the molecular engineering of a light-switching excimer beacon probe to monitor RNase H activity. RNase H is a ribonuclease which can selectively degrade the RNA strand in a RNA–DNA hybrid to produce 3′-hydroxyl- and 5′-phosphate-terminated products without destroying the DNA strand. As shown in Fig. 8.14, binding of the dual-labeled DNA beacon to RNA opens up the hairpin structure and, thus, spatially separates the pyrene moieties. The RNA–DNA hybrid, which serves as the substrate for RNase H cleavage, has a low fluorescence background at 485 nm. After the addition of the enzyme, only the RNA strand will be cleaved from the duplex, which liberates the DNA beacon. The restoration of the hairpin structure brings the pyrene moieties back together and gives a dramatic fluorescence enhancement with excimer emission at 485 nm [25].

8.6 Conclusion and Prospective

Excimer molecular beacons are dual-pyrene-labeled hairpin DNA structures for different targets in chemical and biological systems. With their high quantum yield, large Stokes shift, long fluorescence lifetime, EMBs are becoming increasingly attractive for various applications. Many efforts have been made to improve their design and extend their scope of application. In this chapter, we introduced the design, synthesis, and properties of EMBs and focused on the amplification for different analytes.

Compared to traditional MBs, EMBs possess light-switching signal transduction to minimize background. More importantly, the long fluorescence lifetime holds promising potential applications in real biological samples. Yet, their in vivo application remains challenging, as the near-UV excitation (~340 nm) might be unfavorable for in vivo detection based on insufficient light penetration or possible damage to cells. To solve these problems, appropriate modification of the pyrene moiety might be able to obtain red-shifted excitation pyrene probes. Chemical modification of the probe backbone is an alternative strategy to make it stable in vivo. Finally, coupling the probes with some targeted molecules can reduce undesirable localization, and combining EMB with nanotechnology might also be a promising research direction.

References

1. Tyagi S, Kramer FR (1996) Molecular beacons: probes that fluoresce upon hybridization. Nat Biotechnol 14:303–308
2. Tyagi S, Bratu DP, Kramer FR (1998) Multicolor molecular beacons for allele discrimination. Nat Biotechnol 16:49–53
3. Tyagi S, Marras SAE, Kramer FR (2000) Wavelength-shifting molecular beacons. Nat Biotechnol 18:1191–1196
4. Zhang P, Beck T, Tan W (2001) Design of a molecular beacon DNA probe with two fluorophores. Angew Chem Int Ed 40:402–405
5. Wang K, Tang Z, Yang C et al (2009) Molecular engineering of DNA: molecular beacons. Angew Chem Int Ed 48:856–870
6. Dubertret B, Calame M, Libchaber AJ (2001) Single-mismatch detection using gold-quenched fluorescent oligonucleotides. Nat Biotechnol 19:365–370
7. Song S, Liang Z, Zhang J et al (2009) Gold-nanoparticle-based multicolor nanobeacons for sequence-specific DNA analysis. Angew Chem Int Ed 48:8670–8674
8. Jayagopal A, Halfpenny KC, Perez JW et al (2010) Hairpin DNA-functionalized gold colloids for the imaging of mRNA in live cells. J Am Chem Soc 132:9789–9796
9. Fujimoto K, Shimizu H, Inouye M (2004) Unambiguous detection of target DNAs by excimer-monomer switching molecular beacons. J Org Chem 69:3271–3275
10. Yamana K, Ohshita Y, Fukunaga Y et al (2008) Bis-pyrene-labeled molecular beacon: a monomer-excimer switching probe for the detection of DNA base alteration. Bioorg Med Chem 16:78–83
11. Conlon P, Yang C, Wu Y et al (2008) Pyrene excimer signaling molecular beacons for probing nucleic acids. J Am Chem Soc 130:336–342
12. Yamana K, Iwase R, Furutani S et al (1999) 2′-pyrene modified oligonucleotide provides a highly sensitive fluorescent probe of RNA. Nucleic Acids Res 27:2387–2392
13. Wilson JN, Kool ET (2006) Fluorescent DNA base replacements: reporters and sensors for biological systems. Org Biomol Chem 4:4265–4274
14. Paris PL, Langenhan JM, Kool ET (1998) Probing DNA sequences in solution with a monomer-excimer fluorescence color change. Nucleic Acids Res 26:3789–3793
15. Matray TJ, Kool ET (1999) A specific partner for abasic damage in DNA. Nature 399:704–708
16. Marti AA, Li X, Jockusch S et al (2006) Pyrene binary probes for unambiguous detection of mRNA using time-resolved fluorescence spectroscopy. Nucleic Acids Res 34:3161–3168
17. Huang J, Zhu Z, Bamrungsap S et al (2010) Competition-mediated pyrene-switching aptasensor: probing lysozyme in human serum with a monomer-excimer fluorescence switch. Anal Chem 82:10158–10163
18. Huang J, Wu Y, Chen Y et al (2011) Pyrene-excimer probes based on the hybridization chain reaction for the detection of nucleic acids in complex biological fluids. Angew Chem Int Ed 50:401–404

19. Yang C, Jockusch S, Vicens M et al (2005) Light-switching excimer probes for rapid protein monitoring in complex biological fluids. Proc Natl Acad Sci U S A 102:17278–17283
20. Wang C, Wu C, Chen Y et al (2011) Pyrene excimer for DNA sensors. Curr Org Chem 15:465–476
21. Häner R, Biner SM, Langenegger SM et al (2010) A highly sensitive, excimer-controlled molecular beacon. Angew Chem Int Ed 49:1227–1230
22. Nagatoishi S, Nojima T, Juskowiak B et al (2005) A pyrene-labeled G-quadruplex oligonucleotide as a fluorescent probe for potassium ion detection in biological applications. Angew Chem Int Ed 44:5067–5070
23. Zheng J, Li J, Gao X et al (2010) Modulating molecular level space proximity: a simple and efficient strategy to design structured DNA probes. Anal Chem 82:3914–3921
24. Zheng J, Li J, Jiang Y et al (2011) Design of aptamer-based sensing platform using triple-helix molecular switch. Anal Chem 83:6586–6592
25. Chen Y, Yang C, Wu Y et al (2008) Light-switching excimer beacon assays for ribonuclease H kinetic study. Chembiochem 9:355–359

Chapter 9
Strategies for Delivering Molecular Beacons into Cells

Zehui Cao

Abstract Because of their inherent advantages of detection without separation, high sensitivity, and excellent selectivity, MBs are ideal probes for direct imaging of mRNA in complex cellular environments. However, in order for MBs to function inside cells, methods have to be developed for efficient delivery of MBs into cells. In this chapter, some of the most common techniques for delivering MBs into living cells for RNA detection are reviewed. The advantages and limitations of each technique are discussed.

9.1 Introduction

Molecular beacons are particularly indispensable in intracellular RNA imaging and tracking [1, 2]. Prior to the emergence of MBs, imaging of native mRNA inside living cells was nearly impossible because no probes were able to separate real signals from background signals. Molecular beacons, with their unique hairpin and fluorophore-quencher design, generate fluorescent signals only when bound to their nucleic acid targets. This unique feature, coupled with the unparalleled selectivity of MBs for targets, makes these probes especially suitable for direct imaging of mRNA in complex cellular environments. Messenger RNA has long been linked to biological conditions and functions of cells. Variations of expression levels of mRNAs can often be indications of changes in the cellular system and, ultimately, can even be used for disease diagnosis. For example, DNA microarray technology uses probes immobilized on a DNA chip to analyze expression of a large number of mRNAs, enabling clinicians to compare the results between patients and healthy subjects to identify genes linked to certain diseases [3]. Similarly, studying and imaging the localization and movement of mRNAs inside living cells are important

---

Z. Cao (✉)
Dow AgroSciences LLC, Indianapolis, IN, USA
e-mail: zcao@dow.com

C.J. Yang and W. Tan (eds.), *Molecular Beacons*, DOI 10.1007/978-3-642-39109-5_9,

for understanding the mechanisms of mRNA synthesis, trafficking, and functions, and how diseases such as cancers, as well as potential drug molecules, can affect these behaviors. Direct RNA imaging also helps explain how RNA viruses infect and function in human cells [4]. Recent discoveries have put small interfering RNA (siRNA) and microRNA (miRNA), as well as their regulatory roles in cells, in the spotlight. MB-based RNA imaging can potentially be further explored to study these new classes of RNA molecules and their intracellular behavior and functions [5].

Recently, MBs have been used for RNA expression analysis in conjunction with flow cytometry [6–8], in which MBs hybridize to intracellular RNA while a flow cytometer is employed to measure MB-generated fluorescence in individual cells. Unlike expression measurements from extracted RNAs in bulk solutions, this MB/flow cytometry approach offers RNA detection at the single-cell level, and it allows analysis of cell-to-cell expression variations in a statistically meaningful fashion. This additional information can be uniquely useful for studying how cells at different stages, or from different subpopulations, react to disease conditions, drug molecules, and environmental stimuli.

Despite the great promise of MBs as intracellular RNA probes, one obstacle must be overcome before MBs can even interact with their RNA targets. Specifically, the cellular membrane forms a natural barrier against charged macromolecules, including nucleic acid-based MBs. Efficient delivery methods to overcome this barrier have to be available for any MB-based intracellular studies. Accordingly, strategies for delivering MBs into living cells continue to be explored. Most of these approaches have been focused on direct delivery to cell cytoplasm, instead of endocytosis-based pathways. This preference is first based on the fact that mRNA is mainly located in cytoplasm. Second, endocytosis-based internalization will leave MBs trapped in the endosomes, which are known to contain acidic solutions and nucleases that can both degrade MBs. Some of the major techniques reported for MB delivery include the use of microinjection, microporation, reversible permeabilization, membrane-penetrating peptides, and nanomaterials. These different approaches will be discussed and compared in this chapter.

9.2 Microinjection

Microinjection was the main delivery method in the early days of MB-based RNA imaging, most likely because it was the most straightforward approach. This method involves a glass micropipette with a fine tip that usually has a diameter of 1 μm. A micromanipulator is installed on a microscope stage and is used to bring the micropipette tip close to the cell of interest with high precision under the microscope. Once the glass tip touches the cell, a small pressure is applied through the tip which disrupts the cell membrane and delivers the MB probe solution in the micropipette into the cell. Microinjection has several advantages; most importantly, it allows for the immediate monitoring of the cell for the response of the probe. Secondly, it delivers a relatively reproducible amount of probe to the experimental cell of choice. Gewirtz and colleagues were among the first to image mRNA in single living cells using microinjected MBs [9]. They detected higher fluorescent signals from target cells than from microinjected control cells. However, they also observed degradation of MBs after the MBs stayed in the cells for an extended time. Tan and coworkers reported imaging of abundant beta-actin mRNA in PtK2 cells using microinjected MBs [10]. Another report from the same group described quantitative measurements of cell-to-cell variation in the expression of manganese superoxide dismutase (MnSOD) mRNA by using an MB-based ratiometric imaging approach [11]. Realizing the problem of MB degradation inside living cells, Tyagi and coworkers developed nuclease-resistant molecular beacons for imaging the transport and localization of oskar mRNA in *Drosophila melanogaster* oocytes [12]. Several reports discovered that the microinjected MBs tended to accumulate inside the cell nucleus within minutes after injection, instead of remaining in the cytoplasm for mRNA detection [13, 14]. The mechanism by which the MBs enter the cell nucleus was proposed to be an active facilitated transportation [14]. To solve this problem, MBs were linked to a bulky protein, streptavidin, through a biotin attachment on the MB [13, 14] or covalently linked to a transfer RNA (tRNA) [15]. These modifications did indeed result in reduced sequestration of MBs in nuclei, making mRNA imaging in cytoplasm more practical. In a related study, MBs were used to track diffusion of mRNA-protein complexes inside the cell nucleus [16].

Although simple in principle, the microinjection-based delivery approach has some major limitations. First, it requires costly special equipment for injection tip manipulation on the microscope. Second, proficient equipment operation and cell injection may take prolonged training and practice, thus limiting the type of personnel who can carry out desired studies. In addition, cells can only be injected one at a time, leading to limited numbers of cells that can be analyzed each time. The difficulties associated with microinjection operation also mean poor reproducibility that could result in significant cell-to-cell variations. Overall, studying more than a few cells simultaneously using MBs is impractical with microinjection.

9.3 Microporation

Electroporation is a widely used technique for delivering various molecules, including nucleic acids, into living cells. It is based on the increased cell membrane permeability caused by an external electrical field [17]. When given electrical pulses of certain voltage and duration, it is generally believed that cell plasma membranes begin to form nanometer-scale pores that allow diffusion of external molecules into cells. For different cell types, optimum voltage and pulse duration need to be experimentally determined to achieve the best delivery efficiency. Voltages over a certain limit will lead to irreversible electroporation and reduced cell viability. Electroporation is relatively easy to carry out and usually involves placing cells to be transfected in a special cuvette with two electrodes connected to a high-voltage generator. In this way, large numbers of cells can be transfected at the same

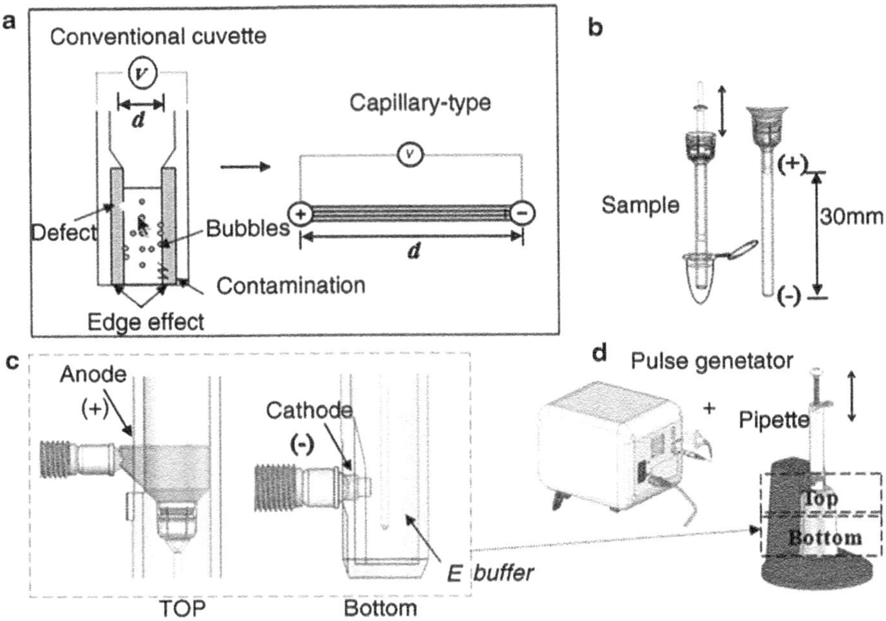

Fig. 9.1 (**a**) Comparison of electrochemical side effects resulting from the physical geometry of two parallel plate-type electrodes integrated in the cuvette system and a single wire-type electrode used in the capillary system; (**b**) schematic of the piston-driven capillary tip (diameter 0.65 mm, length 30 mm); (**c**) the position of the anode and cathode in the capillary electroporation system; (**d**) schematic of the overall capillary electroporation system, which consists of a high-voltage pulse generator and a pipette station (Reprinted from Ref. [23], Copyright 2008, with permission from Elsevier)

time. Electroporation is highly efficient in introducing nucleic acids into mammalian cells. This capability has resulted in its wide application in gene therapy [18], DNA vaccination [19], mRNA and siRNA delivery [20, 21], and plant transformation [22].

Despite its success in nucleic acid delivery, electroporation is often linked to substantial cell damage, with typical cell viability at 20–50 % after electroporation [23], even after careful determination of optimum conditions through tedious experiments. To address this problem, a capillary-based electroporation system was developed (Fig. 9.1) [23]. Termed microporation, this technique uses a thin metal wire inserted in a capillary as the anode. After the mixture of cells and the reagents to be delivered are pulled up into the capillary, the other end of the capillary is inserted in a buffer covering the cathode. Finally, electrical pulses generated by a voltage generator are applied for cell membrane pore formation and cargo delivery inside the capillary. Compared to traditional electroporation, the authors noted that this new design significantly reduced changes in buffer composition, pH, and other factors during electroporation that might have contributed to cell death. In addition to increased viability, cells were much more tolerant to changes in electrical parameters, meaning that less experimental effort is required to determine optimum conditions. More importantly, using this approach, consistently high transfection rates were achieved, even for those cell types considered hard to transfect.

Tsourkas and coworkers tested cytosolic delivery of NeutrAvidin-conjugated MB probes using microporation and other methods [7]. They observed lysosome entrapment with a NeutrAvidin-fluorophore conjugate with all delivery methods. After pegylation of NeutrAvidin conjugates, only microporation produced uniform cytosolic distribution with high transfection efficiency and cell viability. It was found that this uniform fluorescent signal could be retained inside the cytoplasm for at least 24 h, making temporal monitoring of mRNA levels possible. The authors also noted that microporation needed fewer cell samples compared to other methods, such as Streptolysin O (SL-O)-based delivery. When cells were microporated using pegylated MB-NeutrAvidin conjugates and analyzed with a flow cytometer, up to five times increase in sensitivity was achieved compared to unconjugated MBs, as a result of the improved signal-to-background ratios. Based on these findings, the same group used microporation delivery to compare 2′-*O*-methyl (2Me)- and 2′-*O*-methyl-phosphorothioate (2MePS)-modified MBs conjugated to NeutrAvidin for their intracellular trafficking, false-positive signal generation, and functionality as mRNA imaging probes [24]. Tsourkas and coworkers later designed a new imaging probe called the ratiometric bimolecular beacon (RBMB) and used microporation to study its intracellular performance. The RBMB probe consisted of a conventional MB domain and a siRNA-like domain used to facilitate nuclear export. The siRNA domain was designed to replace proteins like streptavidin and quantum dots previously used to keep MBs from entering the cell nucleus. The RBMB was found to remain in the cell cytoplasm, which led to reduced false-positive signals and significantly higher signal-to-background ratios compared to conventional MBs.

9.4 Reversible Permeabilization

Similar to electroporation, reverse permeabilization also delivers the MBs through pores created in the cell membrane. However, the two techniques differ in the way they create the pores. Electroporation can create pores in the cell membrane through the application of an electrical pulse to the cells. Reverse permeabilization involves the use of the chemical Streptolysin O (SL-O) to create pores in the cell membrane. In both methods, loss of materials from inside the cell can occur through the pores, and the amount of probes delivered into each cell can vary. Reverse permeabilization also requires time to form the pores and then additional time to allow the pores to reseal.

SL-O is a pore-forming toxin secreted by a class of spherical Gram-positive bacteria called *Streptococcus*. SL-O binds to cholesterol on the cell membrane and then oligomerizes into ring-shaped structures with pore sizes of approximately 25–30 nm in diameter [25]. This increases the permeability of cell membranes to macromolecules, including protein and nucleic acids. Pore formation by SL-O can be a reversible process, which is achieved by substituting SL-O containing serum-free media after SL-O treatment of cells with regular growth media containing serum [26]. For delivery of nucleic acids, they are simply incubated with SL-O and the target cells during SL-O treatment. This simplicity and reversibility make SL-O attractive as a routine method for delivering oligonucleotides into living cells. For example, it has been widely used for intracellular delivery of antisense oligonucleotides for gene regulation studies [27, 28]. Spiller and coworkers compared delivery of oligodeoxyribonucleotide (ODN) using SL-O, electroporation, and lipophilic conjugation [29]. They found SL-O and electroporation to provide significantly higher delivery efficiency than the lipophilic conjugation-based approach, with SL-O delivering the highest ODN concentrations into cells. However, the authors also noted that some cells showed resistance to permeabilization at low SL-O concentrations. While increasing SL-O concentration improved permeabilization and uniformity, decreased target specificity was observed by the substantial ODN concentration delivered to the cells.

A number of reports have taken advantage of SL-O-based delivery of oligonucleotides for intracellular imaging of RNA using MBs. Bao and coworkers designed a dual-MB approach for mRNA imaging, where two MBs were constructed to bind side by side at the same location of the mRNA sequence [30]. When they bound to the target mRNA, the two fluorophores, each from one MB, were close enough to generate FRET (fluorescence resonance energy transfer) signals for target detection. This approach alleviated false-positives with the single-MB approach based on opening by protein binding or nuclease digestion. The authors used SL-O to deliver MBs into living cells and found surprisingly little fluorescence from cell nucleus, even when they delivered a linear oligonucleotide probe, contradicting previous reports. They explained this result by the low probe concentrations used in the experiments. The same research group later used SL-O to deliver MBs for viral RNA detection in living cells and observed very high delivery efficiency [4]. Following these reports, SL-O was also employed for MB delivery to study mRNA accessibility, detect virus mRNA, and investigate the relationship between MB backbone modification and mRNA translational stage [31–33].

The SL-O-based delivery system has been shown to be simple and efficient for MB imaging applications. Notably, it can transfect a large number of cells at the same time and does not require special equipment for the transfection, making it a more convenient method over other techniques, including microinjection and microporation, and especially suitable for routine and high-throughput intracellular MB delivery. Nonetheless, several factors need to be considered when choosing SL-O as the delivery method. First, as pointed out by Santangelo [25], SL-O binds to cholesterol, and each cell type may have a different membrane cholesterol composition. Consequently, the degree of SL-O-induced membrane permeability may vary between cell types. This would call for the optimization of SL-O transfection protocols for each new cell type to achieve comparable delivery efficiency. Second, standard SL-O protocols require incubation of MB with SL-O and cells for about 30 min, after which the cells are washed and incubated with regular media for another 15 to 60 min before fluorescence imaging. Therefore, to achieve reliable

RNA imaging, necessary modification strategies need to be applied to the MBs to keep them inside cytoplasm and avoid nonspecific interactions with cellular proteins and degradation by nucleases during incubation. Similarly, imaging applications where temporal resolution is critical or rapid reactions need to be monitored may find the SL-O-based delivery method problematic.

9.5 Cell-Penetrating Peptides

Cell-penetrating peptides (CPPs) are a group of peptides that can carry a variety of cargos through the cell membrane. The most widely used CPPs are the arginine-rich cationic peptides derived from HIV Tat protein, with a typical core peptide sequence of RKKRRQRRR. In 1994, Fawell and coworkers reported that conjugation of Tat peptides to a group of proteins led to efficient cellular internalization of these proteins [34]. Since then, CPPs, especially Tat peptides, have been adopted to deliver a wide range of molecules and cargos into living cells, including proteins and peptides [35, 36], small molecules [37, 38], nucleic acids including siRNA and DNA plasmid [39, 40], liposomes, quantum dots, and other nanomaterials [41, 42]. How Tat peptides penetrate the cell membrane is still a debatable topic, with different groups using different cell/cargo systems often reporting contradicting results [43–45]. Many different models of cell penetration have been proposed based on these divergent experimental results, including endocytosis mediated by clathrin or caveolae formation, macropinocytosis, and direct translocation by electrostatic interaction between cationic CPPs and anionic phospholipid head groups of the membrane bilayer [45]. More evidence is emerging that CPPs can take multiple routes into cells, depending on the cell system and cargo they carry.

Tat peptides were first used for MB delivery by Bao and coworkers [46]. Three strategies were designed to covalently link a Tat peptide to the MBs (Fig. 9.2). In all three designs, a modified dT nucleotide that carried a carbon-12 linker was placed at the third base from the 3′-end on the quencher arm of the stem. The end of the linker had a biotin group for cross-linking with a biotin-attached Tat peptide through a streptavidin and a thiol group for directly linking to a maleimide-modified peptide or for connecting to a cysteine-modified peptide via formation of a disulfide bond. Interestingly, the authors found little impact of the positively charged peptide on the performance of all three types of MB/Tat conjugates in buffer solutions. Compared to unconjugated MBs, the MB/Tat complexes had similar kinetics and signal-to-background (S/B) ratios upon addition of excess targets, with only streptavidin-biotin-linked MB/Tat showing slightly slower kinetics and lower S/B ratio. These MB/Tat conjugates were then incubated with cells at 37 °C for 30 min before washing and visualizing. True mRNA detection and localization in the cells were obtained, while false-positive signal was observed using commercial transfection systems. This peptide-based MB delivery system seemed to have circumvented the endocytic pathway, as evidenced by the lack of MB degradation. The authors also noted that the use of 2′-*O*-methyl-modified MBs was not necessary in this study. In

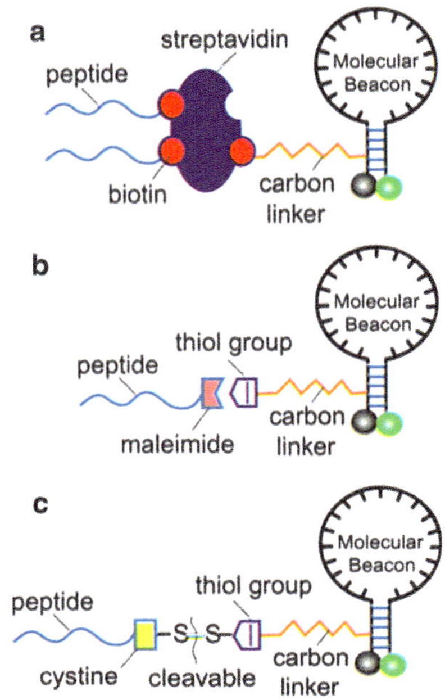

Fig. 9.2 A schematic illustration of three different conjugation schemes for linking the delivery peptide to molecular beacons (Reprinted from Ref. [46] by permission of Oxford University Press)

addition, no nucleus accumulation of MB was mentioned as a significant problem in this report. In another report, Bao and coworkers used peptide-linked MBs for mRNA detection and found that both mRNA targets colocalized with mitochondria [47]. Chen and colleagues used a similar approach to link a Tat peptide to a nuclease-resistant MB and applied it for imaging of Coxsackievirus B6 [48]. They were able to monitor virus spreading among cells in real time. In a related report, Bao and coworkers covalently linked a single nuclear localization sequence (NLS) peptide to an MB with natural bases and utilized it for RNA imaging inside the cell nuclei [49]. SL-O was used to deliver MBs into the cells, while NLS peptide carried the MBs into nuclei. Interestingly, they found that the NLS peptide was needed to bring MBs into nuclei, while unconjugated MBs mostly stayed in cytoplasm.

Peptide-based MB delivery is relatively simple, fast, and efficient and requires no special equipment. From the limited number of reports, nucleus accumulation and degradation of MBs do not seem to be significant concerns, which may have been a result of the peptide attachment. However, an extra step of covalent conjugation is required to make the MB-peptide complex, even though maleimide-thiol chemistry is a relatively simple and fast reaction. In addition, extra care should be taken when choosing MB-peptide for RNA localization studies inside living cells. Since the internalization mechanism of CPP is seemingly complex with many possible routes, the peptide itself can often be distributed to and concentrated in certain locations of the intracellular space, for example, the endosomes. Therefore, localization of MB-peptide complex may not always reflect real distribution of the mRNA targets.

9.6 Gold Nanoparticles

Based on their unique features, gold nanoparticles (AuNPs) have emerged in recent years as a new intracellular delivery platform [50, 51]. First, the gold core is generally considered inert and nontoxic, making it compatible with most biological specimens. Second, the sizes of AuNPs can be fine-tuned with great uniformity in a wide diameter range from a few nanometers to micrometers. AuNPs in the low nanometer range provide very high surface area-to-volume ratios that are capable of accommodating a large quantity of cargo molecules for increased delivery efficiency. Even more important, the surface of AuNPs can be easily functionalized through various physical and chemical interactions. Cargo molecules can therefore be directly linked to AuNPs, or a second layer can be created on the AuNP surface to convey new surface properties that are more accessible to certain cargo molecules. In an early report, plasmid DNA was directly coated on AuNPs 1–3 μM in size, most likely by physical adsorption, and delivered to rabbit skin cells [52]. After that, subsequent studies first coated AuNPs with polyethylenimine (PEI) [53, 54], a highly positively charged polymer that is commonly used for DNA transfection. Plasmid DNA was then mixed to interact with the PEI layer via electrostatic forces during which time DNA condensation could also take place. Neutralized and condensed DNA was more readily taken up by cells. These studies observed higher delivery efficiency and lower cytotoxicity with the AuNP delivery system compared to that with PEI alone. AuNP-assisted oligonucleotide delivery was explored by Mirkin and coworkers. Monothiol- and tetrathiol-modified antisense oligonucleotides were covalently immobilized on 13 nm AuNPs. Unexpected efficient uptake of these DNA-AuNPs throughout the cytoplasm was observed for all tested cell types, even in the absence of any positively charged polymers, such as PEI. Higher gene downregulation and lower cytotoxicity were obtained compared to commercial cell transfection kits. A later report from the same group investigated the mechanism behind this cellular uptake and found that the DNA-coated AuNPs readily bound to up to 23 serum proteins per particle after exposure to cell media, judging by increased particle sizes and decreased negative charges [55]. In addition, the internalization capability seemed to correlate with the number of proteins on the particles.

To take advantage of AuNP-based internalization for intracellular mRNA imaging, Mirkin and coworkers prepared a new probing agent called "nanoflares" (Fig. 9.3) [56], in which a short oligonucleotide-recognizing target mRNA sequence was immobilized on AuNPs. A fluorophore-labeled reporter strand hybridized with the recognition strand such that the fluorophore was close to the gold surface. This design took advantage of the fact that gold is an excellent quencher to fluorophores in close proximity [57]. When the nanoflares were taken up by cells, the target mRNA could bind to the recognition strand and, hence, release the reporter strand. Fluorescence of the fluorophore on the released strand would then be restored, producing a detectable signal. The authors reported lower background fluorescence from nanoflares compared to conventional MBs, and they attributed this finding to

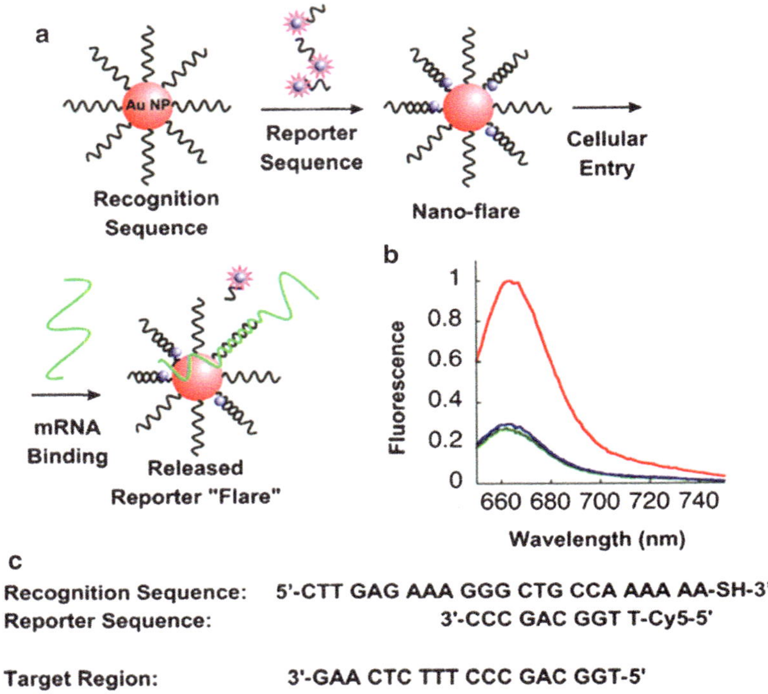

Fig. 9.3 Schematics of nanoflares (Reprinted with the permission from Ref. [56]. Copyright 2007 American Chemical Society)

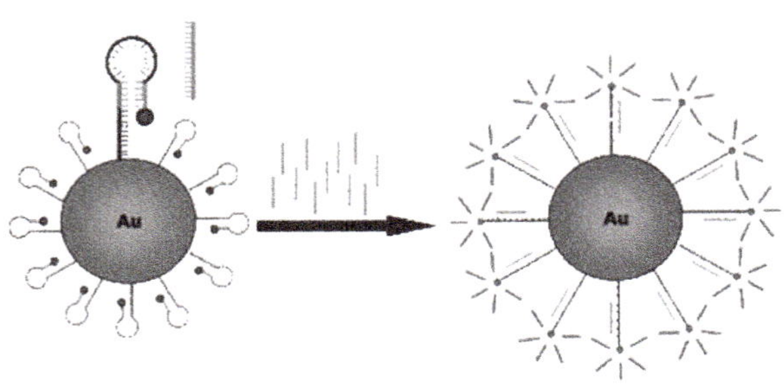

Fig. 9.4 Schematics of hAuNP (Reprinted with the permission from Ref. [59]. Copyright 2010 American Chemical Society)

gold's superior quenching efficiency and protection of surface-bound DNA against nuclease degradation. Wright and colleagues reported 15 nm AuNPs immobilized with an MB-like hairpin DNA (hAuNP) (Fig. 9.4) [58]. One stem of the hairpin DNA was directly linked to the AuNP surface, while the other stem possessed a fluorophore. In the absence of targets, the hairpin structure maintained close proximity between the fluorophore and the gold surface, leading to quenched fluorescence. This hAuNP was readily internalized by SK-MEL-28 cells where the target tyrosinase mRNA hybridized to the hairpin DNA and extended the fluorophore farther away from the AuNP surface, causing an increased fluorescence signal. Specific detection of target mRNA was achieved in SK-MEL-28 cells, but not in control cells. In another report, the same group applied a similar hAuNP design to the imaging of cellular mRNA and viral RNA [59]. In an effort to elucidate the underlying mechanism of nonspecific internalization of DNA-coated AuNPs, the authors evaluated a series of inhibitors of classical endocytic pathways, inhibitors of cell metabolism, such as sodium azide/2-deoxy-D-glucose, and 4 °C pretreatment for their impact on hAuNP internalization in serum-free cell media. Interestingly, none of the treatments significantly affected hAuNP internalization, indicating that a nonendocytosis or at least nonclassical endocytosis model played a major role. Additionally, the presence of serum in the media actually lowered hAuNP internalization, seemingly contradicting the previous report [55].

Taken together, it is clear that more comprehensive studies are needed to confirm the real mechanism of spontaneous DNA-AuNPs internalization. Nonetheless, evidence available so far suggests that AuNPs are capable of delivering DNA probes nonspecifically and very efficiently into the cytoplasm of living cells. Probes immobilized on AuNPs tend to have lower background and good protection against nuclease degradation [60], which are two critically important properties for intracellular imaging by nucleic acid probes. Even though an extra linking step is required to make the AuNP-nucleic acid complex, the thiol-Au conjugation chemistry involved is remarkably simple and convenient. One downside of the current AuNP delivery technique is the long incubation time needed for AuNPs to enter cells, usually hours or longer. However, the resistance of AuNPs to nucleases and their tendency to stay in cytoplasm may compensate for this problem in many applications. Nonetheless, applications requiring monitoring of initial MB-RNA interactions may need to resort to other faster delivery techniques, such as microporation. Moreover, the size and properties of the AuNPs could be major factors in determining the movements of the AuNP-MB complex inside the cells, thus possibly producing unreliable RNA localization and trafficking information.

9.7 Conclusion

In summary, some of the most common techniques for delivering MBs into living cells for RNA detection were reviewed and compared. Selection of a proper delivery approach is the first critical step for MB-based intracellular analysis, and considerations for making such selection include the objectives of the project at hand and the intended application, possible impact of the specific delivery method, as well as cost and available resources.

References

1. Santangelo P, Nitin N, Bao G (2006) Nanostructured probes for RNA detection in living cells. Ann Biomed Eng 34(1):39–50
2. Tyagi S (2009) Imaging intracellular RNA distribution and dynamics in living cells. Nat Methods 6(5):331–338
3. Sorensen KD, Orntoft TF (2010) Discovery of prostate cancer biomarkers by microarray gene expression profiling. Expert Rev Mol Diagn 10(1):49–64
4. Santangelo P, Nitin N, LaConte L, Woolums A, Bao G (2006) Live-cell characterization and analysis of a clinical isolate of bovine respiratory syncytial virus, using molecular beacons. J Virol 80(2):682–688
5. Persat A, Santiago JG (2011) MicroRNA profiling by simultaneous selective isotachophoresis and hybridization with molecular beacons. Anal Chem 83(6):2310–2316
6. Horejsh D, Martini F, Poccia F, Ippolito G, Di Caro A, Capobianchi MR (2005) A molecular beacon, bead-based assay for the detection of nucleic acids by flow cytometry. Nucleic Acids Res 33(2):e13
7. Chen AK, Behlke MA, Tsourkas A (2008) Efficient cytosolic delivery of molecular beacon conjugates and flow cytometric analysis of target RNA. Nucleic Acids Res 36(12):e69
8. King FW, Liszewski W, Ritner C, Bernstein HS (2011) High-throughput tracking of pluripotent human embryonic stem cells with dual fluorescence resonance energy transfer molecular beacons. Stem Cells Dev 20(3):475–484
9. Sokol DL, Zhang X, Lu P, Gewirtz AM (1998) Real time detection of DNA.RNA hybridization in living cells. Proc Natl Acad Sci U S A 95(20):11538–11543
10. Perlette J, Tan W (2001) Real-time monitoring of intracellular mRNA hybridization inside single living cells. Anal Chem 73(22):5544–5550
11. Drake TJ, Medley CD, Sen A, Rogers RJ, Tan W (2005) Stochasticity of manganese superoxide dismutase mRNA expression in breast carcinoma cells by molecular beacon imaging. Chembiochem Eur J Chem Biol 6(11):2041–2047
12. Bratu DP, Cha BJ, Mhlanga MM, Kramer FR, Tyagi S (2003) Visualizing the distribution and transport of mRNAs in living cells. Proc Natl Acad Sci U S A 100(23):13308–13313
13. Tsuji A, Koshimoto H, Sato Y, Hirano M, Sei-Iida Y, Kondo S, Ishibashi K (2000) Direct observation of specific messenger RNA in a single living cell under a fluorescence microscope. Biophys J 78(6):3260–3274
14. Tyagi S, Alsmadi O (2004) Imaging native beta-actin mRNA in motile fibroblasts. Biophys J 87(6):4153–4162
15. Mhlanga MM, Vargas DY, Fung CW, Kramer FR, Tyagi S (2005) tRNA-linked molecular beacons for imaging mRNAs in the cytoplasm of living cells. Nucleic Acids Res 33(6):1902–1912
16. Vargas DY, Raj A, Marras SA, Kramer FR, Tyagi S (2005) Mechanism of mRNA transport in the nucleus. Proc Natl Acad Sci U S A 102(47):17008–17013
17. Neumann E, Schaefer-Ridder M, Wang Y, Hofschneider PH (1982) Gene transfer into mouse lyoma cells by electroporation in high electric fields. EMBO J 1(7):841–845
18. Wells DJ (2004) Gene therapy progress and prospects: electroporation and other physical methods. Gene Ther 11(18):1363–1369
19. Sardesai NY, Weiner DB (2011) Electroporation delivery of DNA vaccines: prospects for success. Curr Opin Immunol 23(3):421–429
20. Tavernier G, Andries O, Demeester J, Sanders NN, De Smedt SC, Rejman J (2011) mRNA as gene therapeutic: how to control protein expression. J Control Release 150(3):238–247
21. Calegari F, Marzesco AM, Kittler R, Buchholz F, Huttner WB (2004) Tissue-specific RNA interference in post-implantation mouse embryos using directional electroporation and whole embryo culture. Differentiation 72(2–3):92–102
22. Rakotondrafara AM, Jackson JR, Kneller EP, Miller WA (2007) Preparation and electroporation of oat protoplasts from cell suspension culture. Current protocols in microbiology. John Wiley & Sons, Inc. Online ISBN: 9780471729259

23. Kim JA, Cho K, Shin MS, Lee WG, Jung N, Chung C, Chang JK (2008) A novel electroporation method using a capillary and wire-type electrode. Biosens Bioelectron 23(9):1353–1360
24. Chen AK, Behlke MA, Tsourkas A (2009) Sub-cellular trafficking and functionality of 2′-O-methyl and 2′-O-methyl-phosphorothioate molecular beacons. Nucleic Acids Res 37(22):e149
25. Santangelo PJ (2010) Molecular beacons and related probes for intracellular RNA imaging. Wiley Interdiscip Rev Nanomed Nanobiotechnol 2(1):11–19
26. Barry EL, Gesek FA, Friedman PA (1993) Introduction of antisense oligonucleotides into cells by permeabilization with streptolysin O. Biotechniques 15(6):1016–1018, 1020
27. Giles RV, Ruddell CJ, Spiller DG, Green JA, Tidd DM (1995) Single base discrimination for ribonuclease H-dependent antisense effects within intact human leukaemia cells. Nucleic Acids Res 23(6):954–961
28. Clark RE, Grzybowski J, Broughton CM, Pender NT, Spiller DG, Brammer CG, Giles RV, Tidd DM (1999) Clinical use of streptolysin-O to facilitate antisense oligodeoxyribonucleotide delivery for purging autografts in chronic myeloid leukaemia. Bone Marrow Transplant 23(12):1303–1308
29. Spiller DG, Giles RV, Grzybowski J, Tidd DM, Clark RE (1998) Improving the intracellular delivery and molecular efficacy of antisense oligonucleotides in chronic myeloid leukemia cells: a comparison of streptolysin-O permeabilization, electroporation, and lipophilic conjugation. Blood 91(12):4738–4746
30. Santangelo PJ, Nix B, Tsourkas A, Bao G (2004) Dual FRET molecular beacons for mRNA detection in living cells. Nucleic Acids Res 32(6):e57
31. Rhee WJ, Santangelo PJ, Jo H, Bao G (2008) Target accessibility and signal specificity in live-cell detection of BMP-4 mRNA using molecular beacons. Nucleic Acids Res 36(5):e30
32. Wang W, Cui ZQ, Han H, Zhang ZP, Wei HP, Zhou YF, Chen Z, Zhang XE (2008) Imaging and characterizing influenza A virus mRNA transport in living cells. Nucleic Acids Res 36(15):4913–4928
33. Nitin N, Rhee WJ, Bao G (2009) Translation inhibition reveals interaction of 2′-deoxy and 2′-O-methyl molecular beacons with mRNA targets in living cells. Nucleic Acids Res 37(15):4977–4986
34. Fawell S, Seery J, Daikh Y, Moore C, Chen LL, Pepinsky B, Barsoum J (1994) Tat-mediated delivery of heterologous proteins into cells. Proc Natl Acad Sci U S A 91(2):664–668
35. Silhol M, Tyagi M, Giacca M, Lebleu B, Vives E (2002) Different mechanisms for cellular internalization of the HIV-1 Tat-derived cell penetrating peptide and recombinant proteins fused to Tat. European J Biochem/FEBS 269(2):494–501
36. Liang JF, Yang VC (2005) Insulin-cell penetrating peptide hybrids with improved intestinal absorption efficiency. Biochem Biophys Res Commun 335(3):734–738
37. Rousselle C, Clair P, Lefauconnier JM, Kaczorek M, Scherrmann JM, Temsamani J (2000) New advances in the transport of doxorubicin through the blood–brain barrier by a peptide vector-mediated strategy. Mol Pharmacol 57(4):679–686
38. Lindgren M, Rosenthal-Aizman K, Saar K, Eiriksdottir E, Jiang Y, Sassian M, Ostlund P, Hallbrink M, Langel U (2006) Overcoming methotrexate resistance in breast cancer tumour cells by the use of a new cell-penetrating peptide. Biochem Pharmacol 71(4):416–425
39. Chiu YL, Ali A, Chu CY, Cao H, Rana TM (2004) Visualizing a correlation between siRNA localization, cellular uptake, and RNAi in living cells. Chem Biol 11(8):1165–1175
40. Ignatovich IA, Dizhe EB, Pavlotskaya AV, Akifiev BN, Burov SV, Orlov SV, Perevozchikov AP (2003) Complexes of plasmid DNA with basic domain 47–57 of the HIV-1 Tat protein are transferred to mammalian cells by endocytosis-mediated pathways. J Biol Chem 278(43):42625–42636
41. Yagi N, Yano Y, Hatanaka K, Yokoyama Y, Okuno H (2007) Synthesis and evaluation of a novel lipid-peptide conjugate for functionalized liposome. Bioorg Med Chem Lett 17(9):2590–2593
42. Santra S, Yang H, Stanley JT, Holloway PH, Moudgil BM, Walter G, Mericle RA (2005) Rapid and effective labeling of brain tissue using TAT-conjugated CdS:Mn/ZnS quantum dots. Chem Commun (Camb) 25:3144–3146

43. Brooks H, Lebleu B, Vives E (2005) Tat peptide-mediated cellular delivery: back to basics. Adv Drug Deliv Rev 57(4):559–577
44. Futaki S (2005) Membrane-permeable arginine-rich peptides and the translocation mechanisms. Adv Drug Deliv Rev 57(4):547–558
45. Sawant R, Torchilin V (2010) Intracellular transduction using cell-penetrating peptides. Mol Biosyst 6(4):628–640
46. Nitin N, Santangelo PJ, Kim G, Nie S, Bao G (2004) Peptide-linked molecular beacons for efficient delivery and rapid mRNA detection in living cells. Nucleic Acids Res 32(6):e58
47. Santangelo PJ, Nitin N, Bao G (2005) Direct visualization of mRNA colocalization with mitochondria in living cells using molecular beacons. J Biomed Opt 10(4):44025
48. Yeh HY, Yates MV, Mulchandani A, Chen W (2008) Visualizing the dynamics of viral replication in living cells via Tat peptide delivery of nuclease-resistant molecular beacons. Proc Natl Acad Sci U S A 105(45):17522–17525
49. Nitin N, Bao G (2008) NLS peptide conjugated molecular beacons for visualizing nuclear RNA in living cells. Bioconjug Chem 19(11):2205–2211
50. Alkilany AM, Murphy CJ (2010) Toxicity and cellular uptake of gold nanoparticles: what we have learned so far? J Nanopart Res: Iinterdisciplin Forum Nanoscale Sci Technol 12(7):2313–2333
51. Rana S, Bajaj A, Mout R, Rotello VM (2011) Monolayer coated gold nanoparticles for delivery applications. Adv Drug Deliv Rev 64:200–216
52. Sundaram P, Xiao W, Brandsma JL (1996) Particle-mediated delivery of recombinant expression vectors to rabbit skin induces high-titered polyclonal antisera (and circumvents purification of a protein immunogen). Nucleic Acids Res 24(7):1375–1377
53. Thomas M, Klibanov AM (2003) Conjugation to gold nanoparticles enhances polyethylenimine's transfer of plasmid DNA into mammalian cells. Proc Natl Acad Sci U S A 100(16):9138–9143
54. Ow Sullivan MM, Green JJ, Przybycien TM (2003) Development of a novel gene delivery scaffold utilizing colloidal gold-polyethylenimine conjugates for DNA condensation. Gene Ther 10(22):1882–1890
55. Giljohann DA, Seferos DS, Patel PC, Millstone JE, Rosi NL, Mirkin CA (2007) Oligonucleotide loading determines cellular uptake of DNA-modified gold nanoparticles. Nano Lett 7(12):3818–3821
56. Seferos DS, Giljohann DA, Hill HD, Prigodich AE, Mirkin CA (2007) Nano-flares: probes for transfection and mRNA detection in living cells. J Am Chem Soc 129(50):15477–15479
57. Dubertret B, Calame M, Libchaber AJ (2001) Single-mismatch detection using gold-quenched fluorescent oligonucleotides. Nat Biotechnol 19(4):365–370
58. Harry SR, Hicks DJ, Amiri KI, Wright DW (2010) Hairpin DNA coated gold nanoparticles as intracellular mRNA probes for the detection of tyrosinase gene expression in melanoma cells. Chem Commun (Camb) 46(30):5557–5559
59. Jayagopal A, Halfpenny KC, Perez JW, Wright DW (2010) Hairpin DNA-functionalized gold colloids for the imaging of mRNA in live cells. J Am Chem Soc 132(28):9789–9796
60. Seferos DS, Prigodich AE, Giljohann DA, Patel PC, Mirkin CA (2009) Polyvalent DNA nanoparticle conjugates stabilize nucleic acids. Nano Lett 9(1):308–311

Chapter 10
Molecular Beacons for Intracellular Analysis: Challenges and Successes

Colin D. Medley and Zhi Zhu

Abstract In this chapter, the design of MB for intracellular imaging application is detailed. Applications of molecular beacons for intracellular analysis are reviewed. Challenges and solutions for intracellular applications of MBs are discussed.

10.1 Introduction

Since their development in 1996, molecular beacons (MBs) have been used in a variety of roles, including qPCR [1], identification of mycobacterial species [2], protein study [3–5], single-base mismatch detection [6–8], and incorporation in biosensors and DNA arrays [9–12]. Of the multitude of applications using MBs, intracellular monitoring is one of the most technically demanding for both the analyst and the probe itself. In the cytoplasm of a cell, MBs find themselves in a rather inhospitable environment, including pH gradients that can impact fluorophores, enzymes that can digest the probe, and a small abundance of complementary sequences in a vast amount of mRNA. Despite these challenges, MBs are the ideal probe for intracellular use as a result of their ability to detect complementary sequences without separating out unbound probes. This enables the detection of real-time mRNA expression inside of a living cell, which can be a powerful tool for understanding biological mechanisms and studying how different conditions can impact the cell.

C.D. Medley (✉)
Small Molecule Pharmaceutical Sciences, Genentech, USA
e-mail: medley.colin@gene.com

Z. Zhu
Department of Chemical Biology, College of Chemistry and Chemical Engineering, Xiamen University, Xiamen 361005, P.R. China
e-mail: zhuzhi@xmu.edu.cn

C.J. Yang and W. Tan (eds.), *Molecular Beacons*, DOI 10.1007/978-3-642-39109-5_10,

Given the complexity of the intracellular environment, several factors must be taken into consideration when using MBs inside living cells. The first challenge is concentration. Excessive amounts of probe create a fluorescence background that makes it difficult to detect changes in the fluorescence level associated with target hybridization. On the other hand, too few probes make the detection of the fluorescence more challenging and risk saturation of the probes such that differences in expression levels are not detected. Another challenge is consistency. If variation in the amount of probes delivered between cells is too high, analysis of the mRNA expression inside the cell can be compromised. Once inside the cell, the MB must now survive enzymatic cleavage from DNases and nonspecific protein binding from single-strand binding protein that melts the stem of the probe. Both the binding and digestion of the probe result in the restoration of fluorescence, mimicking a target hybridization event. Despite these issues, the utility of MBs has been successfully demonstrated inside living cells, leading to many developments that improve both the way probes are used in the intracellular matrix and the way probes are designed and constructed. Many technological advances, such as ratiometric analysis and the inclusion of nonnative nucleic acid bases into the probes, have enabled the quantitative analysis of mRNA expression on a single-cell level.

10.2 Molecular Beacon Design for Intracellular RNA Detection

The use of MBs for intracellular RNA detection and localization requires the design of the MB such that it recognizes mRNA inside a cell. The design of the MBs with this special purpose involves three major areas: the loop, the stem, and the fluorophore/quencher pair. The loop region is of primary concern when designing MBs for intracellular use. The first step requires the selection of an appropriate target region on the mRNA sequence for the MB, particularly since large mRNA sequences exhibit a complex secondary structure. The MB must be able to reach to its complementary sequence through the secondary structure, or the MB will not hybridize and produce the fluorescent signal for detection. Current mRNA folding programs do not give a completely reliable secondary structure; thus, a series of MBs is generally selected for each mRNA sequence. Once possible target sequences are identified through the RNA folding programs, the complementary loop sequence can be designed. After the loop region is designed, the stem region is next and is constructed so that the stem region is not complementary to any part of the loop. Self-complementary regions between the loop itself and the loop with stem can lead to stable secondary structures in the probe that greatly increase the background fluorescence. Another effective strategy is incorporating base pairs in the stem that also hybridize to the target sequence. This yields a larger region of complementary bases that increases the stability of the hybrid state and reduces the phenomenon of sticky-end base pairing that can reduce sensitivity.

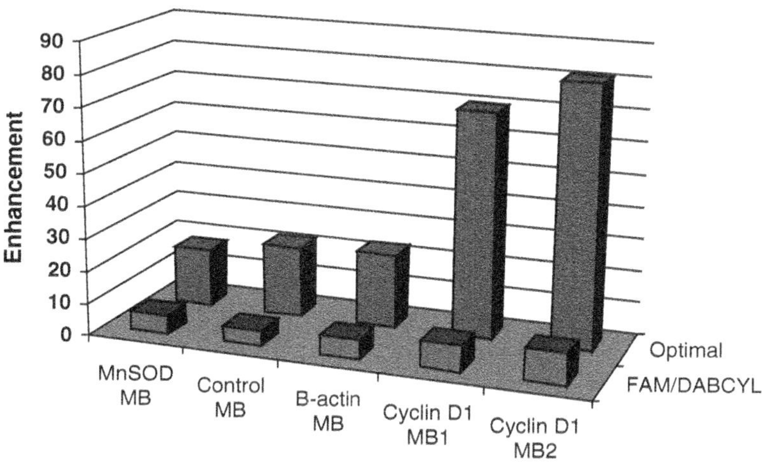

Fig. 10.1 Comparison of the enhancements of MBs with FAM/Dabcyl and optimized fluorophore/quencher pairs

Another more practical concern when designing and synthesizing MBs for any application is the final nucleotide base positioned before the fluorophore. Several investigators have observed that nucleotides can quench the fluorescence emission of fluorophores with guanine (G) to be the best quencher among the bases [13–15]. As a result, taken the advantage of this property, MBs designed with a C connecting the fluorophore and a G linking the quencher moiety display better signal enhancement than other arrangements.

Once the DNA portion of the MB is designed, the fluorophore and quencher pair can be chosen. The proper selection of the fluorophore and quencher pair is another important factor [16]. This is generally based on the instrumentation available for the experiment to ensure efficient excitation and detection of the fluorophore in the open state, as well as high quenching efficiency in the closed state. While many probes are initially tested with FAM/Dabcyl fluorophore/quencher pairs, these are often replaced for the actual applications to achieve better sensitivities. Figure 10.1 shows the fluorescence enhancements of several MBs using FAM/Dabcyl pairs and then the higher quality pairs used in subsequent applications. In these experiments, only the fluorophore/quencher pair was changed, and each MB was at a concentration of 1 μM with a target concentration of 10 μM. The optimal fluorophore/quencher pairs are Alexa Fluor 647 and Blackhole Quencher 3 for the MnSOD MB, Alexa Fluor 555 and Blackhole Quencher 2 for the control MB, Alexa Fluor 488 and Blackhole Quencher 1 for the B-actin MB, and Cy3 and Blackhole Quencher 2 for both of the Cyclin D1 MBs. The complete sequences and fluorophore/quencher pairs are listed in Table 10.1. In each case, the optimized fluorophore/quencher pair exhibits both a great signal enhancement between the open and closed states and a greater overall level of fluorescence. Both of these improvements are especially useful when dealing with the small number of probes and target sequences in intracellular applications.

Table 10.1 Molecular beacon sequences

MB name	Sequence with fluorophore/quencher pair
Control MB1	5′-Cy5-CCT AGC TCT AAA TCG CTA TGG TCG CGC TAG G-BHQ3-3′
Control MB2	5′-AF555-CCT AGC TCT AAA TCG CTA TGG TCG CGC TAG G-BHQ2-3′
B-actin MB1	5'-TMR-CCG TCG AGG AAG GAA GGC TGG AAG AGC GAC GG-BHQ2-3′
B-actin MB2	5′-AF488-CCG TCG AGG AAG GAA GGC TGG AAG AGC GAC GG-BHQ1-3′
Cyclin D1 MB1	5′-Cy3-ACG ACG GCC ACC ACG CTC CCC GCT GCC ACC GTC GT-BHQ2-3′
Cyclin D1 MB2	5′-Cy3-GCA GCA TCC AGG TGG CGA CGA TCT TGC TGC-BHQ2-3′
MnSOD MB	5′-AF647-CCG AGC CAG TTA CAT TCT CCC AGT TGA TTG CTC GG-BHQ3-3′

10.3 Molecular Beacon Delivery

Delivery of MBs inside the cell has been an area of considerable research effort, which has resulted in many effective options for intracellular delivery. The most common delivery methods include microinjection [17], electroporation [18], peptide-assisted delivery [19], and reversible permeabilization [20]. The detailed introduction and technical description can refer to Chap. 9. Here we provide several examples of using these methods for DNA intracellular delivery to compare their delivery efficient and required time sale.

10.3.1 Electroporation

Electroporation is a common method for gene transfection and can refer to either single-cell electroporation or the mass delivery of materials to a whole population of cells. The electroporation method essentially creates pores in the cell membrane through the application of an electrical pulse to the cells. The pulse causes the cell membrane to develop pores which allows any probes surrounding the cells to pass through the cell membrane. The major advantage of electroporation is its ability to achieve a mass delivery of the probes with a very short (less than 1 min) incubation time, thus allowing real-time monitoring of the probe. In this experiment, electroporation, as a mass delivery method, was performed using a BTX ECM 830 pulse generator (Harvard Apparatus) with a PP35-2P Petri Pulser electrode. The electrode, with an electrical pulse sufficient to deliver DNA into the cell, is designed to fit a 35 mm cell culture dish. The initial settings were based on a protocol for MDA-MB-231 cells provided by BTX, using a pulse of 300 V for 50 μs with a 1 μM concentration of fluorophore-labeled DNA.

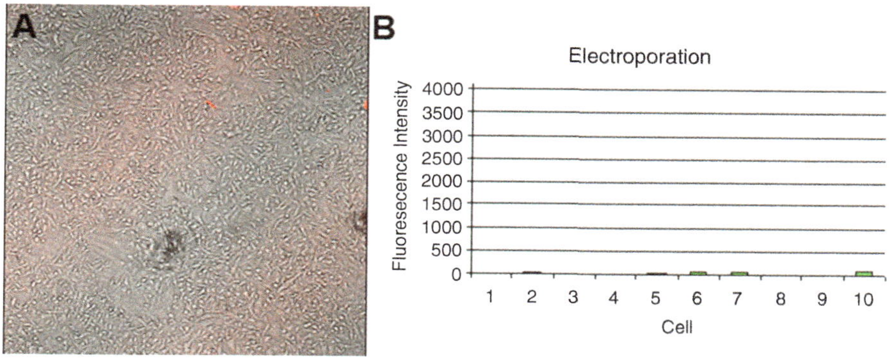

Fig. 10.2 (**a**) Fluorescence images of cells after electroporation in a 5 μM DNA-spiked medium. (**b**) Average cell intensities of ten cells after electroporation

This protocol produced no noticeable change in the fluorescence of the cells. The voltage, pulse length, concentration of DNA, and number of pulses were then increased until reaching the instrument maximums. The results of a representative experiment performed with five pulses of 1,000 V for 50 μs each in Fig. 10.2 with the fluorescence image (A) and the average cell intensities (B) indicated little to no uptake of the DNA transpired. These experiments were repeated using trypsinized cells in BTX electroporation cuvettes, and the cells were imaged after replating in a 35 mm culture dish. However, no fluorescence signal reached levels suitable for use as an MB delivery method. While electroporation functions well for gene transfection, in which only low copy numbers of the genes need to be delivered, these results indicate that the method does not translate well for MB probe delivery. For reliable detection of gene expression, a large amount of probes needs to be delivered to the cells. The failure of the electroporation delivery method means that obtaining high-throughput delivery of the probes for real-time measurements may not be feasible at this juncture, as other methods for mass cellular delivery require incubation times of at least 30 min, and there is no practical method for the high-throughput use of microinjection, which can guarantee the delivery amount but need to be injected to cells individually.

10.3.2 Reversible Permeabilization

Similar to electroporation, reverse permeabilization also delivers the MBs through pores created in the cell membrane. However, it involves the use of the chemical streptolysin O (SL-O) to create pores in the cell membrane. In order to evaluate the use of reversible permeabilization as a delivery method for MBs, two primary criteria were established. The first and most important was the efficiency of probe delivery. If an insufficient amount of probe is delivered to the cell, then the overall sensitivity of the measurement is affected. Also, the amount delivered to the cell

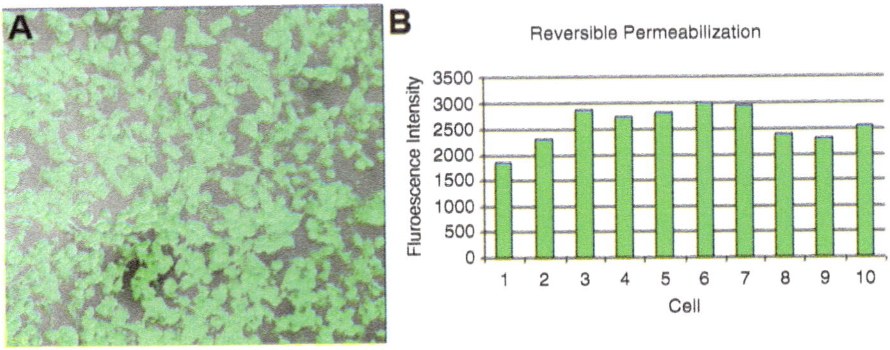

Fig. 10.3 (**a**) Fluorescence image of fluorophore-labeled DNA delivered through reversible permeabilization. (**b**) Average intensities of ten random cells after delivery of probe using reversible permeabilization

needs to be relatively reproducible. There will likely be some cell-to-cell variations in delivery; however, many variations can be compensated by using ratiometric analysis which will be discussed at a later point in the chapter. For probes to be delivered in effective quantities, the second criterion involves an incubation requirement. Since unmodified DNA-based MBs will be used for subsequent experiments, the incubation time for the delivery must take less than 30 min, essentially because degradation of MBs inside the cell will normally occur after this time point. MB degradation also causes the restoration of fluorescence, thereby making it indistinguishable from target hybridization. Therefore, in measurements taken after 30 min, it will be difficult to distinguish which signal resulted from target hybridization and which signal resulted from probe degradation.

Fluorophore-labeled probes were delivered into living cells using a previously published reversible permeabilization protocol [21]. The protocol utilizes SL-O, which binds to cholesterol molecules in the cell membrane to form channels approximately 30 nm in diameter in serum-free media and was shown to be a rapid, efficient, less damaging, and more versatile compared with many conventional transfection methods [22]. Specifically, SL-O was first activated by adding 5 mM of TCEP to 2 U/ml of SL-O for 30 min at 37 °C. Cells grown in 35 mm dishes were incubated for 10 min in 200 ml of serum-free medium containing 0.2 U/ml of activated SL-O (0.5 U SL-O per 10^6 cells) and 1 μM of the Alexa Fluor 488-labeled DNA. The cells were then incubated for different incubation periods of 15, 30, 45, and 60 min. The cells showed uptake of the DNA after 15 min; however, the uptake did not plateau until 45 min (data not shown). Figure 10.3a shows a representative fluorescence image, and Fig. 10.3b shows the average fluorescence intensity of ten random cells. After 45 min, the cells showed a relatively reproducible fluorescence intensity that was quite suitable for imaging or other applications. While the incubation time was too long for conventional DNA MBs, reversible permeabilization will likely be an effective delivery method for probes enhanced with nuclease or protein resistance.

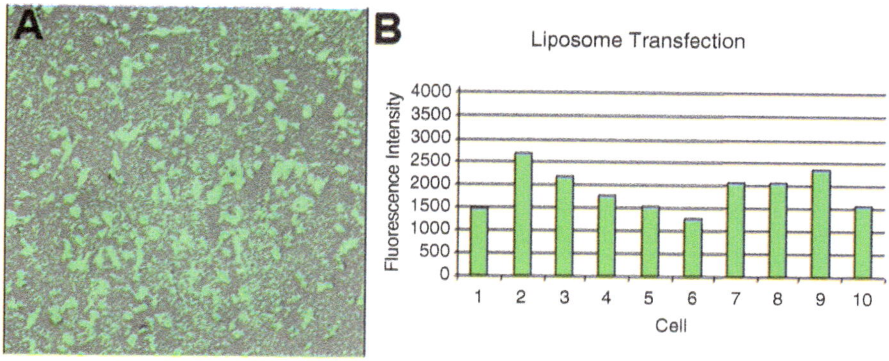

Fig. 10.4 (**a**) Fluorescence image of cells after delivery of fluorophore-labeled DNA using a lipid-based transfection reagent. (**b**) Average fluorescence intensities of ten random cells after delivery of fluorophore-labeled DNA

10.3.3 Liposome Delivery

Lipid-based transfection reagents constitute another common method for the delivery of DNA into cells. These reagents form a bilayer similar to the cell membrane when placed in aqueous environments. Any material stable in the aqueous environment is then trapped inside the membrane. When mixed with the cells, the liposome can deliver their contents inside the cell, either through incorporation of the lipid bilayer into the cell membrane or the breakdown of the bilayer by lysosome inside the cell, both of which cause the release of the materials into the cell. In order to determine whether lipid-based transfection would be suitable for probe delivery, a commercially available transfection reagent Lipofectamine 2000 (Invitrogen) was used to deliver Alexa Fluor 488-labeled DNA. Prior to any experiments, the cells were incubated in antibiotic-free media to prevent any uptake of the antibiotic into the cell. The uptake of the antibiotic into the cell through the liposome complexes is highly toxic to the cells. The stock DNA was then diluted to 1 μM concentration, while 10 μL of Lipofectamine was diluted in a separate 50 μL solution of cell media. Then the diluted Lipofectamine was gently mixed with 50 μL of the DNA solution and incubated for 30 min. The cell media in the cell culture dish were then replaced with the transfection mixture and incubated for 1 h. After the incubation period, the transfection reagents were removed, and the cells were washed three times with cell media. The cells were then imaged, as shown in Fig. 10.4a. The average intensities of ten random cells are plotted in Fig. 10.4b. Incubation times of less than 1 h were also tested and found to have significantly lower fluorescence intensities than that of 1-h incubation. These liposome-based transfection results indicate that this methodology can deliver sufficient amounts of DNA inside the cell for analysis. However, the required incubation times (about 1 h) are unsuitable for DNA-based MBs. It is interesting to note that liposomes appear to have an affinity for the culture dish surface, as well as the cell membrane of the cultured cells. In each experiment conducted with the liposomes, the bare cell culture dish surface generally had several highly fluorescent spots on the surface, even after several washing steps. While this affinity did not appear to affect the overall results, the presence of the spots on the dish can impair the determination of the fluorescence background and lead to some uncertainty in those determinations. In addition, because the liposome-based methods allowed delivery of the MB into the cell without forming artificial pores or physical injection, they had the greatest variability in the delivery, and further use would likely require some form of normalization.

10.3.4 Viral Vector Delivery

Following the establishment of the first infectious clone of AAV serotype 2 (AAV2) in 1982 [23], viral vectors have rapidly gained popularity in gene therapy applications by their lack of pathogenicity, wide range of infectivity, and ability to establish long-term transgene expression [24]. Recombinant AAV2 vectors have been tested in preclinical studies for a variety of diseases, such as hemophilia, 1 anti-trypsin deficiency, cystic fibrosis, Duchenne muscular dystrophy, and rheumatoid arthritis. At least 20 clinical trials have been completed or initiated with 15 different AAV2-based vectors being administered in several hundred patients thus far [24]. Also, AAV vectors have been used for gene delivery in a variety of cell types, including the liver [25–27], muscle [26], brain [28], retina [29], and cancer cells [30]. Therefore, we have sought to determine whether they have potential as a delivery vehicle for DNA-based molecular probes. AAV viral vectors that were modified to express several biotin molecules on their surface were obtained from the labs of Kenneth Warrington. Prior to incubation with the cells, the virus vectors were incubated with excess FITC-labeled streptavidin for 12 h. In theory, the streptavidin would then be used to attach several biotinylated sequences onto the surface of the vector; however, for this proof of concept, the labeled streptavidin alone was used for the quantification of the amount delivered. After incubation with the streptavidin, the viral vectors were incubated with the cells for 1 h, based on published reports at a concentration of 1,000 viral particles per cell. Figure 10.5a depicts a representative fluorescence image of the cells, while Fig. 10.5b shows the average intensities of ten random cells. Overall, the viral vectors delivered a fair amount of material into the cell, although it was less than many of the other methods tested. It should be noted, however, that once any biotinylated probe was added to the vector, 2–3 times as much probe material could be delivered as a consequence of the multiple biotin-binding sites situated on the streptavidin. While further optimization of the delivery protocol would have proved beneficial, there was insufficient virus for subsequent experiments. This indicates still another limitation of the viral vector-based delivery: the lack of commercially available vectors suitable for use with synthetic DNA probes.

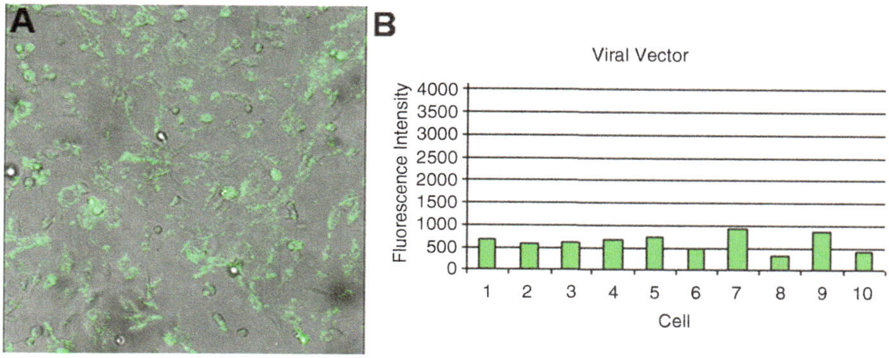

Fig. 10.5 (**a**) Fluorescence image of cells after incubation with a FITC-streptavidin-conjugated viral vector. (**b**) The average intensity of ten random cells after incubation with FITC-streptavidin-linked viral vectors

10.3.5 Microinjection

Microinjection is a well-established technique for the delivery of materials into the cell. The MB solution is loaded into the tips and connected to a microinjector. Once the tip is positioned inside the cell, the microinjector sends a pulse of pressure through the tip that forces the MB solution into the cell. Using microinjection, the primary criterion for the delivery of Alexa Fluor 488-labeled DNA was the amount of DNA delivered. Since no incubation time is necessary for microinjection, it allows the real-time monitoring of hybridization events inside the cell. However, the low throughput and high technical expertise required make systematic studies using microinjection much more difficult and time-consuming. A Leiden microincubator with a TC-202A temperature controller (Harvard Apparatus, Holliston, MA) was used to keep the cells at 37 °C during injection and monitoring. An EXFO Burleigh PCS-6000-150 micromanipulator was used for positioning the injector tip. An Eppendorf Femtojet microinjector with 0.5 μm Femtotips was used to inject the 1 μM Alexa Fluor 488-labeled DNA sequence into the cell. The cells were injected utilizing a 40x 1.35 NA oil immersion objective. The cells were injected with the DNA at 8 psi for a duration of 10 μs.

The cells to be injected were selected based on morphology in an effort to inject healthy cells. Cells displaying a small round morphology or a large bulge-like morphology were not injected. Figure 10.6a shows a representative injected cell, while Fig. 10.6b shows the average intensity from ten different injected cells. Overall, the injected cells had the highest signal intensity of all delivery methods tested and showed good reproducibility. The use of microinjection produces several advantages. Without the need for incubation time, real-time monitoring can proceed. Also, an effective amount of probe can be delivered, and the approach is valid for most adherent cell types.

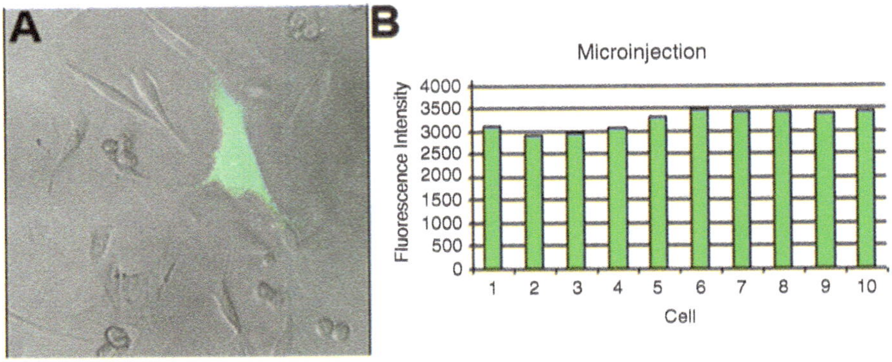

Fig. 10.6 (**a**) A representative image of a cell injected with the fluorophore-labeled DNA. (**b**) The average cell intensities of ten different cells injected with the fluorophore-labeled DNA sequence

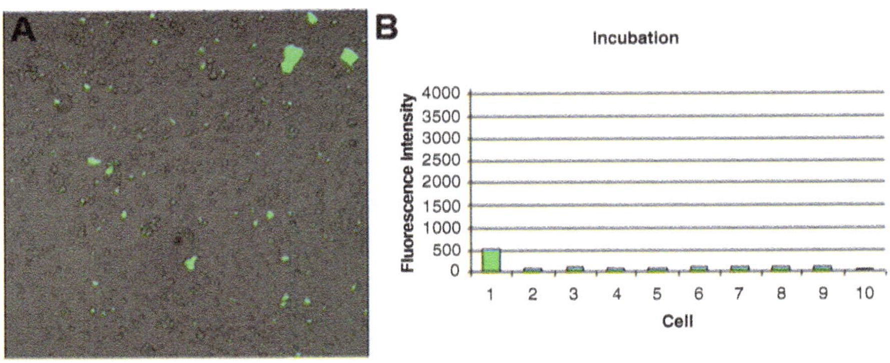

Fig. 10.7 (**a**) Fluorescence image of cells incubated with the fluorophore-labeled DNA sequence for 1 h. (**b**) Plot of the average cell intensities of ten random cells incubated with the fluorophore-labeled DNA

10.3.6 Incubation

The delivery efficiency of a simple incubation of the Alexa Fluor 488-labeled DNA sequence was also performed. This experiment was conducted more as a control for the previous techniques that required an incubation time than as a true delivery method. For these experiments, a 1 μM solution of AF488-labeled DNA was prepared using cell media. The cells were incubated for 1 h in the DNA-spiked cell media, after which the DNA-spiked media were replaced with normal cell media. The fluorescence images and analysis were performed at that time. Figure 10.7a shows the cells after the incubation, while Fig. 10.7b illustrates the average cell intensities of the live cells obtained after incubation. While many of the dead cells (based on morphology) exhibited high fluorescence intensity, indicating an uptake of the probe, living cells showed little to no uptake of the fluorophore-labeled DNA.

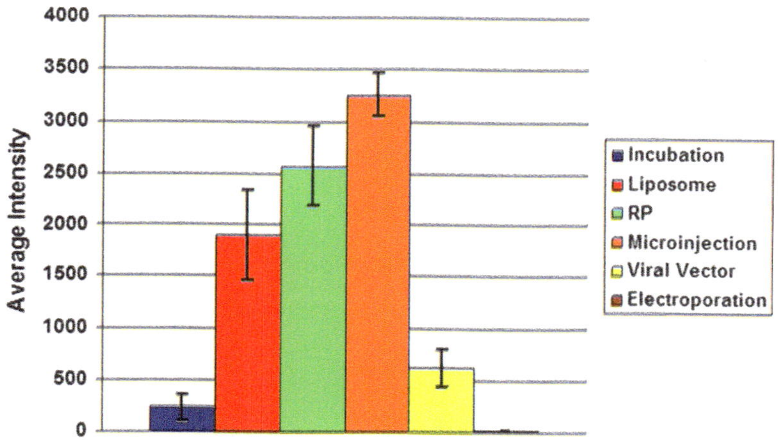

Fig. 10.8 Average intensities from the cells for each DNA delivery method tested

This indicates that incubation alone is insufficient as a delivery method and that the results from the previous experiments where uptake of the DNA occurred did not result from simple incubation of the probe.

10.3.7 Delivery Method Conclusions

In order to more effectively compare the different delivery methods, the average fluorescent intensities of the cells analyzed for each delivery method were calculated and plotted in Fig. 10.8. Microinjection, reversible permeabilization, and the liposome-based transfection all performed well, with each method delivering an experimentally significant amount of the fluorophore-labeled DNA. The viral vector approach shows the potential for being a solid method for MB delivery, although much more work is still required to achieve that goal. Electroporation and incubation were both unsuccessful in delivering the fluorophore-labeled DNA inside the cells. However, in selecting a delivery method for future studies with DNA MBs, there were two main criteria established: that sufficient amounts of material should be delivered and that any incubation must be less than 30 min. The two methods that achieved delivery in less than 30 min were microinjection and electroporation. However, electroporation did not deliver sufficient material for use in the intracellular studies, leaving microinjection as the only viable method left. However, when effective probes that are more stable in the intracellular environment are used, delivery methods, such as liposome transfection and reversible permeabilization, should be very effective options for MB delivery, assuming delivery efficiencies similar to those of the nonstandard MBs. The use of microinjection does create some additional problems like being low throughput and technically demanding, but these

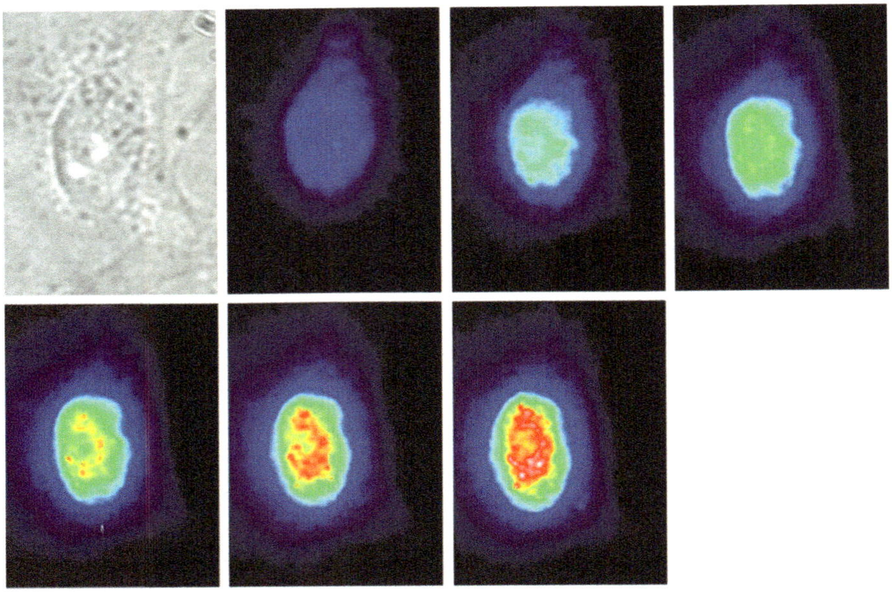

Fig. 10.9 Transmission and the fluorescent images of a PtK2 cell injected with a β-1 adrenergic mRNA MB at 3-min intervals for 18 min (Reprinted with permission from Ref. [31]. Copyright 2001 American Chemical Society)

are inherent to the method and cannot be effectively addressed. The variability that can be seen in the amount of probes delivered and other experimental variations can be addressed through different normalization procedures. In particular, ratiometric analysis can be an effective tool in obtaining more reliable results from even technically demanding intracellular studies.

10.3.8 Applications of Molecular Beacons for Intracellular Analysis

Currently, MBs have been used for intracellular analysis in a variety of cell systems for studying both expression and localization of mRNA. In 2001, the real-time hybridization of an MB to mRNA was visualized inside of a single cell [31]. In this study, MBs specific for β-actin and β-1 adrenergic mRNA were used to detect their respective mRNA targets in kangaroo rat kidney (PtK2) cells. Figure 10.9 shows the sample response of a β-1 adrenergic MB hybridizing to its target inside of a PtK2 cell. The MBs were delivered to the cells using microinjection and then monitored for 18 min. In the PtK2 cells, an increase in fluorescence intensity was detected for both the β-actin and β-1 adrenergic MBs. In order to prove that increase of fluorescence was from the intended interaction, a negative control MB was also monitored for its response inside of a cell. The negative control MB, which had no complement inside the cells, showed no increase in fluorescence inside the cell. This indicates that the fluorescence from the targeted MBs was caused by target-specific hybridization, as opposed to a nonspecific interaction, since any nonspecific interaction or degradation would have equally affected the negative control MB. This study demonstrated the real-time detection capabilities of MBs for mRNA detection and showed conclusively that MBs could be very valuable tools for visualizing gene expression inside single living cells.

While early studies showed the potential of the technology, the variability and uncertainty in the intracellular environment led to more efforts to verify the changes in fluorescence using multiple fluorescent species inside a single cell. In 2003, Tyagi et al. demonstrated that MBs could be used for the visualization of the distribution and transport of mRNA [32]. In this study, an MB for oskar mRNA was investigated in *Drosophila melanogaster* oocytes. They initially demonstrated visualization of the distribution of oskar mRNA in the cell; however, because of the fluorescence background exhibited from the MBs, they decided to use a binary MB approach that utilized two MBs that targeted adjacent positions on the mRNA. When both MBs were hybridized to the mRNA sequence, the donor and acceptor fluorophores were brought within close proximity, thus allowing FRET to occur. This generates a new signal that indicated hybridization of both MBs with the mRNA. In addition to visualizing the mRNA distribution, they were also able to track the migration of the mRNA throughout the cell and even into adjacent cells in the oocyte. This study demonstrated the potential of MBs for studying the localization of mRNA inside of single cells, and it also demonstrated that MBs could be used for tracking mRNA migration, even into different cells. Other studies have been built on this line of investigation and have imaged MBs on viral mRNA inside host cells to study the mRNA behavior [33]. This study investigated the localization of the mRNA inside the cell, and it also utilized photobleaching of the fluorophore on the MB in order to study the diffusion of the MB mRNA hybrids. These studies demonstrate the wealth of information that can be gained through the visualization of MB hybridization inside a single cell.

In addition to localization and distribution, expression levels of mRNA have also been studied inside living cells using MBs. Using a dual-MB FRET approach, the relative expression levels of K-ras and survivin mRNA were determined in human dermal fibroblasts [20]. This involved designing two MBs for adjacent target regions. Once both MBs hybridized to the same mRNA, the fluorophores on the MBs were brought within close proximity, and FRET was allowed to occur. Using this process, they demonstrated the localization of the two mRNA sequences in different cell types and how the different mRNA sequences were localized at different regions inside the cell. The use of the two MBs allowed for much greater specificity since the signal required two separate hybridization events to be generated (Fig. 10.10).

One of the limitations of MBs for intracellular analysis has been the inherent variability in the fluorescent measurements obtained. However, it is difficult to

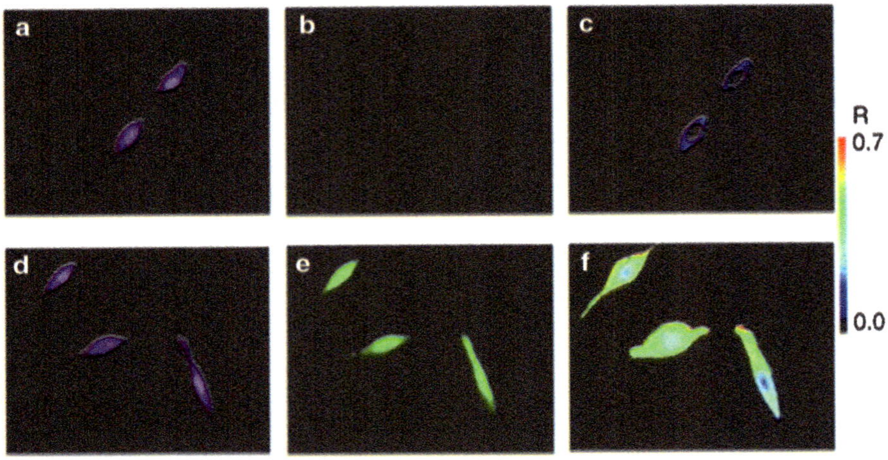

Fig. 10.10 Intracellular imaging of single cells using MB probes. A ratiometric approach has been used to minimize experimental variations and enable more reliable data collection. On the *top row* are the cellular responses for "closed" MBs. On the *bottom row* are the cellular responses for "open" MBs. (**a**) and (**d**) are the fluorescence emission images of a reference probe. (**b**) and (**e**) correspond to fluorescence emission images of the MB probe response. (**c**) and (**f**) are representative ratiometric images of the MB responses by dividing the image from the MB by the image of the reference probe

attribute these variations to any one cause, such as instrument or experimental variability, variations in gene expression, or simply different amounts of probe being delivered into the cell. In an effort to eliminate the experimental and instrumental variations in order to study the stochasticity of gene expression, a method of ratiometric analysis was developed and applied to cancer cell genomics [34]. In this approach, a fluorophore-labeled DNA reference probe was injected along with the MBs. The reference probe exhibited a constant level of fluorescence relative to the MB signal that resulted in a ratiometric value for the MB that was independent of instrumental variations, like the amount of probe delivered to the cell, and experimental variations, like different cell volumes. Since the ratio of the reference probe to the MB is constant, any increase in the ratio would result from the increase of MB fluorescence, indicating target hybridization. The concept of ratiometric analysis is demonstrated in Fig. 10.11. This figure shows that the reference signal remains constant and that differences between hybridized and unhybridized MBs can be easily distinguished.

Using ratiometric analysis to normalize the fluorescence measurements and circumvent the sources of variation, the stochasticity of the gene expression was then studied in detail using cells at basal expression levels, chemical gene stimulation in cells, and cells transfected to overexpress a specific gene. Figure 10.11 shows the stochasticity of the basal gene expression of manganese superoxide dismutase (MnSOD) inside of single cancer cells. As can be seen in Fig. 10.11, the gene of

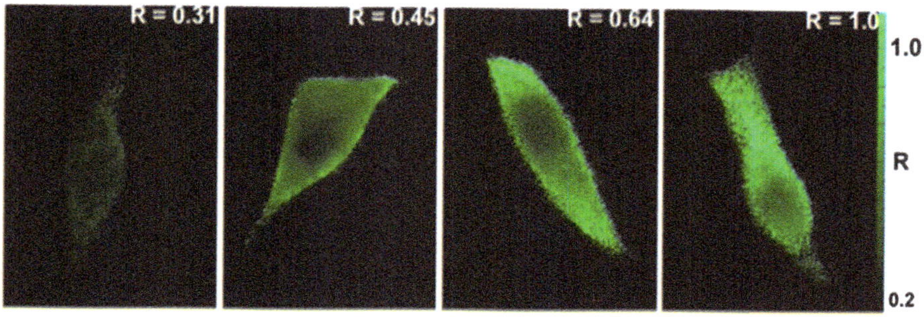

Fig. 10.11 Ratiometric images of MDA-MB-231 cells at 10 min after the injection of MB-2. R values indicate the average ratiometric value of the cytoplasm after analysis

expression of MnSOD varies from cell to cell from very low expression levels to much higher expression levels. Once the ratiometric analysis was developed, the stochasticity of each group of cells was studied for both MnSOD and β-actin mRNA expression using scatterplots. While the β-actin mRNA expression was similar in each group of cells, the MnSOD expression for each group of cells had a distinct pattern. Thus the applicability of the ratiometric analysis was demonstrated and shown to compensate for many of the variations encountered in single-cell analysis.

Other publications have also utilized ratiometric analysis as a means of normalizing the MB signal in order to compare the gene expression profiles in different groups of cells [35]. Here multiple genes were detected simultaneously using ratiometric analysis. In these experiments, β-actin and MnSOD mRNA were detected inside of human breast cancer cells. In addition to MBs for β-actin and MnSOD, control MB with no complement inside of the cell was injected. The MBs were delivered using microinjection and imaged using confocal microscopy. The confocal microscope was able to resolve the three MB fluorophores in addition to the reference probe fluorophore. A sample cell is shown in Fig. 10.12 with the fluorescent images of the MBs and the reference probe. In this figure, it should be noted that (1) the reference probe fluorescence stays constant over the duration of the monitoring period, as the ratiometric analysis assumes that this signal remains constant, and (2) control MB shows a very low level of fluorescence with no increase in signal during the monitoring period. This reinforces that the increase of the fluorescence seen in the β-actin MB channel is due to a specific hybridization event for if the signal increase was due to a nonspecific interaction, the control MB would also be affected.

In order to demonstrate the potential of the technique, the chemical stimulation of gene expression was also studied. This was accomplished by incubating one group of cells with lipopolysaccharide (LPS). LPS was previously shown to increase the expression of MnSOD mRNA. Twenty cells of each type were injected with the mixture of MBs and the reference probe, and the gene expression profile of each cell was determined. The results of this study after ratiometric analysis can be seen in Fig. 10.13. The data in Fig. 10.13 illustrate that the control MB

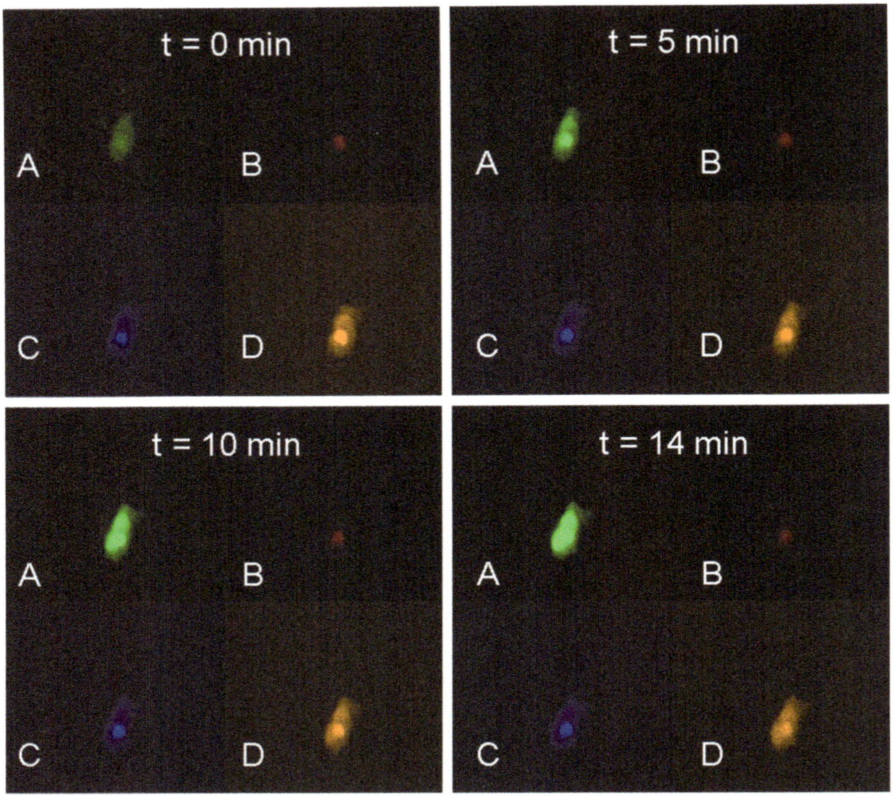

Fig. 10.12 Time-lapse fluorescent images of each MB inside of a single MDA-MB-231 cell. (**a**) The image of the β-actin MB (*green*), (**b**) control MB (*red*), (**c**) MnSOD MB (*blue*), and (**d**) reference probe (*orange*) (Reprinted with permission from Ref. [35]. Copyright 2005 American Chemical Society)

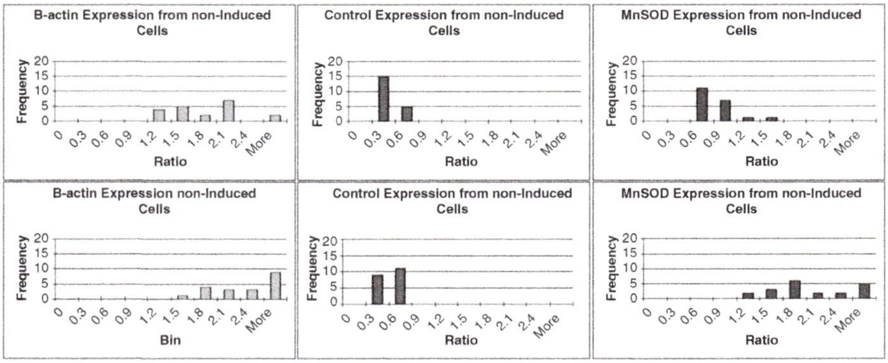

Fig. 10.13 Histograms showing the distribution of ratiometric responses for control and LPS-induced MDA-MB-231 cells with control MB (*middle*), MnSOD (*right*), and B-actin MB (*left*) as measured simultaneously in each cell

signal remains at a low level for both groups of cells. The expression of β-actin also remained very similar between both groups of cells. However, the MnSOD mRNA expression increased dramatically, as can be seen by the shift in the MnSOD expression in the histogram. While the overall trends are consistent, this study also showed considerable variability in gene expression on a cell-to-cell level. This is consistent with the previously mentioned study that also demonstrated a high degree of stochasticity in the gene expression on a cell-to-cell basis. Moreover, these results were consistent with the growing realization that the average cell, as defined by cell population analysis, might not exist [36, 37].

10.4 Improving Molecular Beacon Stability Enables Longer Monitoring Times

Improving the stability of MBs has been the focus of much effort and has led to the incorporation of many types of nonstandard bases into the probes, including locked nucleic acid bases [38], 2′-methoxy RNA [39], phosphorothioated bases [40], and even peptide nucleic acids [41]. The idea behind this approach is to make the MBs resistant to single-strand binding protein and endogenous nucleases that result in limited lifetimes for the probe inside of the cell. LNA-based MBs have already been demonstrated to be more stable inside the cell than DNA-based MBs, as seen in Fig. 10.14.

LNA MBs are limited by their slower hybridization rates which, in turn, impair their effectiveness in real-time monitoring. Further efforts with LNA have focused on chimeric MBs with both DNA and LNA to impart the stability of LNA, along with the hybridization rates of normal DNA [42]. This has resulted in probes that can monitor mRNA expression in real time for hours, as shown in Fig. 10.15 [43].

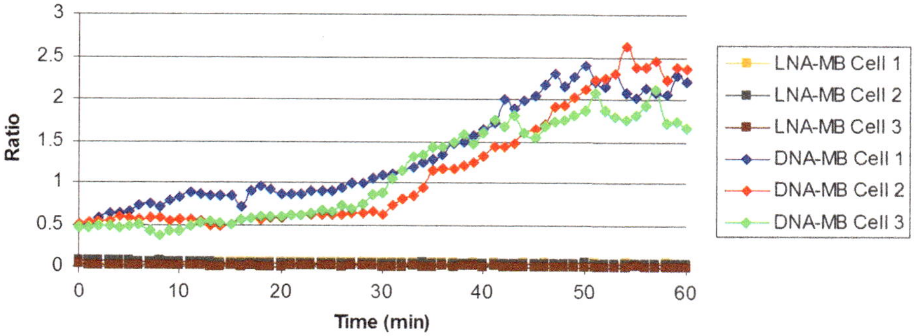

Fig. 10.14 Background signal of LNA- and DNA-based MBs as a function of time inside of single living cells

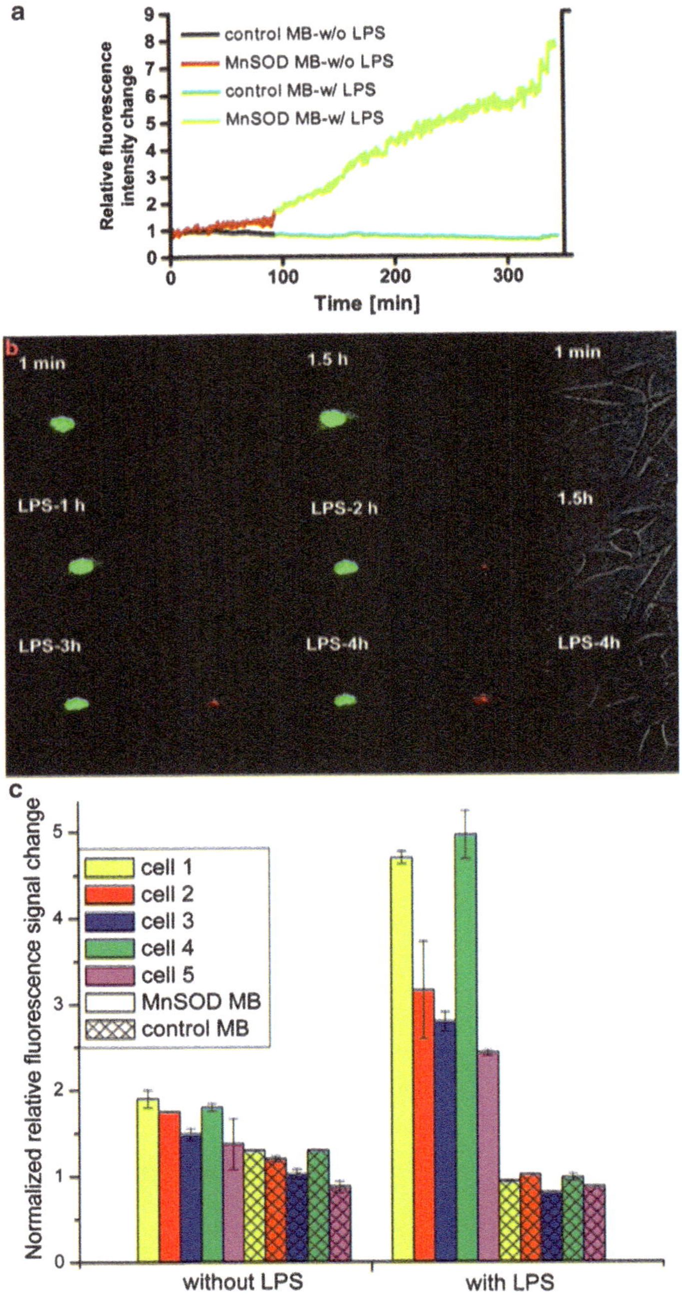
a
control MB-w/o LPS
MnSOD MB-w/o LPS
control MB-w/ LPS
MnSOD MB-w/ LPS
Relative fluorescence intensity change
Time [min]
b
1 min
1.5 h
1 min
LPS-1 h
LPS-2 h
1.5h
LPS-3h
LPS-4h
LPS-4h
c
cell 1
cell 2
cell 3
cell 4
cell 5
MnSOD MB
control MB
Normalized relative fluorescence signal change
without LPS
with LPS

10.5 Intracellular Applications of Molecular Beacons

The major purpose behind the development of advanced techniques and probes is the use of MBs for studying intracellular processes. One such process, the movement of mRNA from translation sites to transcription sites, was visualized using MBs [44]. MBs have also been used to track the movement of ribonucleoprotein complexes through the nucleoplasm to the nuclear periphery [45] and characterize their movement inside of the nucleus [46]. Outside of the nucleus, mRNA has also been visualized colocalizing with mitochondria in the cytoplasm [47]. Many intracellular applications focus on visualizing the localization and movement of mRNA inside of cells. These studies include studying viral mRNA, cancer-related mRNA expression, and mRNA related to DNA damage. Studies have demonstrated monitoring the localization and transport of poliovirus [33] and influenza A [48] viral mRNA inside of individual cells. Inside of host cells, it was found that roughly half of the poliovirus RNA was immobilized inside of the cell, while the other half diffused freely. It was also shown that the revealed entrapment and diffusion mechanism was associated with virus-induced membrane structures. The study of influenza A mRNA transport was able to delve deeper into the transport pathways of viral mRNA utilized in living cells. It found that the export of influenza A virus mRNA is independent of the CRM1 pathway, while the function of RNA polymerase II (RNAP-II) may be needed. Influenza A virus mRNA transport in living cells was characterized, and it was suggested that it may be exported from the nucleus by the cellular TAP/p15 pathway with NS1 protein and RNAP-II participation. Real-time detection of Coxsackievirus replication inside of infected living cells has also been reported using nuclease-resistant MBs [49]. The study of gene expression among cancer-related genes inside single cells has also been a focus of researchers. This includes studying how different treatments impact gene expression [17, 35] and the relative expression levels of specific genes in normal and cancerous cell lines [50]. Exploring the genetic foundations of DNA damage response inside of single cells illustrates a positive correlation between RAD52 gene expression and gamma-ionizing radiation and MMS-induced DNA damage in human TK6 cells [51]. This myriad of applications demonstrate the immense scope of information and understanding that MBs can unlock in the intracellular environment.

Fig. 10.15 Representative microinjection experiments. (**a**) Relative fluorescence signal changes of control MB and MnSOD MB during 1.5 h without LPS treatment and during 4 h with LPS treatment. The *y*-axis represents the signal changes relative to the initial fluorescence signal directly after microinjection (supporting information). (**b**) Time-lapse of control MB (*green*) and MnSOD MB (*red*) inside of a MDA-MB-231 cell. The control MB fluorescence did not change over time, but that for MnSOD changed significantly, as the gene was induced by LPS. (**c**) Histograms of relative fluorescence signal change of MnSOD MBs (no pattern) and control MB (cross-hatched pattern) without LPS and with LPS treatments within different cells (Reprinted with permission from Ref. [43]. Copyright 2008 American Chemical Society)

10.6 Conclusion

Molecular beacons continue to achieve their potential in a variety of applications from DNA arrays to protein studies. One avenue of research that has continued to evolve is the use of MBs for studying living cells. Initial results focused on visualizing the distribution of mRNA expression inside of the cell. This line of inquiry graduated to quantitative application for comparing gene expression over short monitoring times. All the while, many advances in the field of molecular probes, from delivery to composition, continued to push the capabilities of MBs. Currently, multiple MBs are able to remain stable inside of a single cell for hours, enabling research into gene expression pathways and biological mechanisms. With optimized and newly developed tools in hand, future work in this field will focus on using MBs to gain greater understanding of living cells and biological processes from the genetic level.

References

1. Tyagi S, Kramer FR (1996) Molecular beacons: probes that fluoresce upon hybridization. Nat Biotechnol 14:303–308
2. El-Hajj HH, Marras SA, Tyagi S, Shashkina E, Kamboj M, Kiehn TE, Glickman MS, Kramer FR, Alland D (2009) Use of sloppy molecular beacon probes for identification of mycobacterial species. J Clin Microbiol 47:1190–1198
3. Belon C, Frick D (2008) Monitoring helicase activity with molecular beacons. Biotechniques 45:433–442
4. Yang CJ, Li JJ, Tan W (2006) Using molecular beacons for sensitive fluorescence assays of the enzymatic cleavage of nucleic acids. Methods Mol Biol 335:71–81
5. Tang Z, Wang K, Tan W, Li J, Liu L, Guo Q, Meng X, Ma C, Huang S (2003) Real-time monitoring of nucleic acid ligation in homogenous solutions using molecular beacons. Nucleic Acids Res 31:e148
6. Root DD, Vaccaro C, Zhang Z, Castro M (2004) Detection of single nucleotide variations by a hybridization proximity assay based on molecular beacons and luminescence resonance energy transfer. Biopolymers 75:60–70
7. Marras SAE, Kramer FR, Tyagi S (1999) Multiplex detection of single-nucleotide variations using molecular beacons. Genet Anal 14:151–156
8. Wang Y, Wang H, Gao L, Liu H, Lu Z, He N (2005) Polyacrylamide gel film immobilized molecular beacon array for single nucleotide mismatch detection. J Nanosci Nanotechnol 5:653–658
9. Martinez K, Estevez MC, Wu Y, Phillips JA, Medley CD, Tan W (2009) Locked nucleic acid based beacons for surface interaction studies and biosensor development. Anal Chem 81:3448–3454
10. Li J, Tan W, Wang K, Xiao D, Yang X, He X, Tang Z (2001) Ultrasensitive optical DNA biosensor based on surface immobilization of molecular beacon by a bridge structure. Anal Sci 17:1149–1153
11. Fang XH, Liu XJ, Schuster S, Tan WH (1999) Designing a novel molecular beacon for surface-immobilized DNA hybridization studies. J Am Chem Soc 121:2921–2922
12. Yao G, Tan W (2004) Molecular beacon-based array for sensitive DNA analysis. Anal Biochem 331:216–223

13. Crockett AO, Wittwer CT (2001) Fluorescein-labeled oligonucleotides for real-time PCR: using the inherent quenching of deoxyguanosine nucleotides. Anal Biochem 290:89–97
14. Seidel CAM, Schulz A, Sauer MHM (1996) Nucleobase-specific quenching of fluorescent dyes. 1. Nucleobase one-electron redox potentials and their correlation with static and dynamic quenching efficiencies. J Phys Chem 100:5541–5553
15. Tyagi S, Bratu DP, Kramer FR (1998) Multicolor molecular beacons for allele discrimination. Nat Biotechnol 16:49–53
16. Dirks RW, Molenaar C, Tanke HJ (2003) Visualizing RNA molecules inside the nucleus of living cells. Methods 29:51–57
17. Medley CD, Lin H, Mullins H, Rogers RJ, Tan W (2007) Multiplexed detection of ions and mRNA expression in single living cells. Analyst 132:885–891
18. Golzio M, Rols MP, Teissie J (2004) In vitro and in vivo electric field-mediated permeabilization, gene transfer, and expression. Methods 33:126–135
19. Nitin N, Santangelo PJ, Kim G, Nie S, Bao G (2004) Peptide-linked molecular beacons for efficient delivery and rapid mRNA detection in living cells. Nucleic Acids Res 32:e58/1–e58/8
20. Santangelo PJ, Nix B, Tsourkas A, Bao G (2004) Dual FRET molecular beacons for mRNA detection in living cells. Nucleic Acids Res 32:e57/1–e57/9
21. Sokol DL, Zhang X, Lu P, Gewirtz AM (1998) Real time detection of DNA RNA hybridization in living cells. Proc Natl Acad Sci U S A 95:11538
22. Marras SAE, Kramer FR, Tyagi S (2002) Efficiencies of fluorescence resonance energy transfer and contact-mediated quenching in oligonucleotide probes. Nucleic Acids Res 30:e122
23. Carter BJ (2005) Adeno-associated virus vectors in clinical trials. Hum Gene Ther 16:541–550
24. McCarty DM, Monahan PE, Samulski RJ (2001) Self-complementary recombinant adeno-associated virus (scAAV) vectors promote efficient transduction independently of DNA synthesis. Gene Ther 8:1248–1254
25. Wang Z, Ma HI, Li J, Sun L, Zhang J, Xiao X (2003) Rapid and highly efficient transduction by double-stranded adeno-associated virus vectors in vitro and in vivo. Gene Ther 10:2105–2111
26. McCarty DM, Fu H, Monahan PE, Toulson CE, Naik P, Samulski RJ (2003) Adeno-associated virus terminal repeat (TR) mutant generates self-complementary vectors to overcome the rate-limiting step to transduction in vivo. Gene Ther 10:2112–2118
27. Fu H, Muenzer J, Samulski RJ, Breese G, Sifford J, Zeng X, McCarty DM (2003) Self-complementary adeno-associated virus serotype 2 vector: global distribution and broad dispersion of AAV-mediated transgene expression in mouse brain. Mol Ther 8:911–917
28. Yang GS, Schmidt M, Yan Z, Lindbloom JD, Harding TC, Donahue BA, Engelhardt JF, Kotin R, Davidson BL (2002) Virus-mediated transduction of murine retina with adeno-associated virus: effects of viral capsid and genome size. J Virol 76:7651–7660
29. Xu D, McCarty D, Fernandes A, Fisher M, Samulski RJ, Juliano RL (2005) Delivery of MDR1 small interfering RNA by self-complementary recombinant adeno-associated virus vector. Mol Ther 11:523–530
30. Lakowicz JR (ed) (1994) Topics in fluorescence spectroscopy. New York, Plenum Press
31. Perlette J, Tan WH (2001) Real-time monitoring of intracellular mRNA hybridization inside single living cells. Anal Chem 73:5544–5550
32. Bratu DP, Cha BJ, Mhlanga MM, Kramer FR, Tyagi S (2003) Visualizing the distribution and transport of mRNAs in living cells. Proc Natl Acad Sci U S A 100:13308–13313
33. Cui ZQ, Zhang ZP, Zhang XE, Wen JK, Zhou YF, Xie WH (2005) Visualizing the dynamic behavior of poliovirus plus-strand RNA in living host cells. Nucleic Acids Res 33:3245–3252
34. Drake TJ, Medley CD, Rogers RJ, Tan WH (2005) Stochasticity of manganese superoxide dismutase mRNA expression in breast carcinoma cells by molecular beacon imaging. Chembiochem 6:2041–2047
35. Medley CD, Drake TJ, Tomasini JM, Rogers RJ, Tan WH (2005) Simultaneous monitoring of the expression of multiple genes inside of single breast carcinoma cells. Anal Chem 77:4713–4718

36. Chee M, Yang R, Hubbell E, Berno A, Huang XC, Stern D, Winkler J, Lockhart DJ, Morris MS, Fodor SPA (1996) Accessing genetic information with high-density DNA arrays. Science 274:610–614
37. Ginzinger DG (2002) Gene quantification using real-time quantitative PCR. An emerging technology hits the mainstream. Exp Hematol 30:503–512
38. Wang L, Yang CJ, Medley CD, Benner SA, Tan W (2005) Locked nucleic acid molecular beacons. J Am Chem Soc 127:15664–15665
39. Nitin N, Rhee WJ, Bao G (2009) Translation inhibition reveals interaction of 2′-deoxy and 2′-O-methyl molecular beacons with mRNA targets in living cells. Nucleic Acids Res 37:4977–4986
40. Chen AK, Behlke MA, Tsourkas A (2009) Sub-cellular trafficking and functionality of 2′-O-methyl and 2′-O-methyl-phosphorothioate molecular beacons. Nucleic Acids Res 37:e149
41. Ortiz E, Estrada G, Lizardi PM (1998) PNA molecular beacons for rapid detection of PCR amplicons. Mol Cell Probes 12:219–226
42. Yang CJ, Wang L, Wu Y, Kim Y, Medley CD, Lin H, Tan W (2007) Synthesis and investigation of deoxyribonucleic acid/locked nucleic acid chimeric molecular beacons. Nucleic Acids Res 35:4030–4041
43. Wu Y, Yang CJ, Moroz LL, Tan W (2008) Nucleic acid beacons for long-term real-time intracellular monitoring. Anal Chem 80:3025–3028
44. Rodriguez AJ, Condeelis J, Singer RH, Dictenberg JB (2007) Imaging mRNA movement from transcription sites to translation sites. Semin Cell Dev Biol 18:202–208
45. van den Bogaard PT, Tyagi S (2009) Using molecular beacons to study dispersal of mRNPs from the gene locus. Methods Mol Biol 464:91–103
46. Vargas DY, Raj A, Marras SA, Kramer FR, Tyagi S (2005) Mechanism of mRNA transport in the nucleus. Proc Natl Acad Sci U S A 102:17008–17013
47. Santangelo PJ, Nix B, Tsourkas A, Bao G (2004) Dual FRET molecular beacons for mRNA detection in living cells. Nucleic Acids Res 32:e57
48. Wang W, Cui ZQ, Han H, Zhang ZP, Wei HP, Zhou YF, Chen Z, Zhang XE (2008) Imaging and characterizing influenza A virus mRNA transport in living cells. Nucleic Acids Res 36:4913–4928
49. Yeh HY, Yates MV, Mulchandani A, Chen W (2008) Visualizing the dynamics of viral replication in living cells via Tat peptide delivery of nuclease-resistant molecular beacons. Proc Natl Acad Sci U S A 105:17522–17525
50. Peng XH, Cao ZH, Xia JT, Carlson GW, Lewis MM, Wood WC, Yang L (2005) Real-time detection of gene expression in cancer cells using molecular beacon imaging: new strategies for cancer research. Cancer Res 65:1909–1917
51. Riches LC, Lynch AM, Gooderham NJ (2010) A molecular beacon approach to detecting RAD52 expression in response to DNA damage in human cells. Toxicol In Vitro 24:652–660

Chapter 11
Molecular Aptamer Beacons

Cuichen Wu, Chaoyong James Yang, and Weihong Tan

Abstract In this chapter, we present a special class of nucleic acid probes, called molecular aptamer beacons, which combine the specific recognition of aptamers and the signal transduction of molecular beacons. By using SELEX process, single-stranded aptamers can be selected to bind essentially any targets ranging from ions, small molecules, peptides, proteins, virus, and even whole cells. Aptamers have gained increasing attention in the field of molecular recognition, design of biosensors, and investigation of RNAs in living cells due to their good selectivity, high affinity, convenient synthesis, and easy modification. There are different designs of molecular aptamer beacons, such as replacing the loop part of molecular beacons into a target-specific sequence, splitting an aptamer into two subunits that reconstructed in the presence of target molecules, and structure-switching molecular beacons. In addition, we also present detailed discussion on the molecular aptamer beacons developed for fluorescent, electrochemical, and optical applications.

C. Wu (✉)
Department of Chemistry, University of Florida, Gainesville, FL 32611, USA
e-mail: cwu@chem.ufl.edu

C.J. Yang
Department of Chemical Biology, College of Chemistry and Chemical Engineering, Xiamen University, Xiamen, P.R. China
e-mail: cyyang@xmu.eud.cn

W. Tan
Department of Biomedical Engineering and Department of Chemistry, Hunan University, Changsha, P.R. China

Department of Chemistry, University of Florida, Gainesville, FL, USA
e-mail: tan@chem.ufl.edu

C.J. Yang and W. Tan (eds.), *Molecular Beacons*, DOI 10.1007/978-3-642-39109-5_11,

11.1 Introduction

Conventional molecular beacons (MBs), firstly reported by Tyagi and Kramer in 1996 [1], are hairpin-shaped oligonucleotide with fluorescence-donor and fluorescence-acceptor moieties on either terminal of a nucleic acid sequence. The loop sequence is usually designed to be the complementary part of a special target nucleic acid. Due to the specific recognition of target nucleic acid sequences, a spontaneous conformation change occurs upon hybridization to restore the fluorescence. By monitoring the change of fluorescence intensity, the original MBs were used for the detection of DNA and RNA in living systems [1–4], design of biosensors [5, 6], and investigation of nonspecific protein-DNA interaction [7–9]. However, MBs lack sequence specificity in protein recognition, which only limits a few nonspecific ssDNA-binding proteins used in these assays [10, 11].

To overcome the limitations in lacking of specific recognition for MBs, another special class of nucleic acid probes, aptamers, is presented. Aptamers are single-stranded oligonucleotides selected to bind essentially any targets with high selectivity and affinity through a process called SELEX (Systematic Evolution of Ligands by EXponential Enrichment, Fig. 11.1) [13–15]. As the nucleic acid equivalent of antibodies, aptamers have transcended base-recognition ability of nucleic acids to recognize almost every type of species ranging from ions, small molecules, peptides, proteins, virus, and even whole cells. Aptamers can bind to target protein molecules as well as nucleic acids with an excellent selectivity and sensitivity based on their high affinity acquired during the SELEX process. There

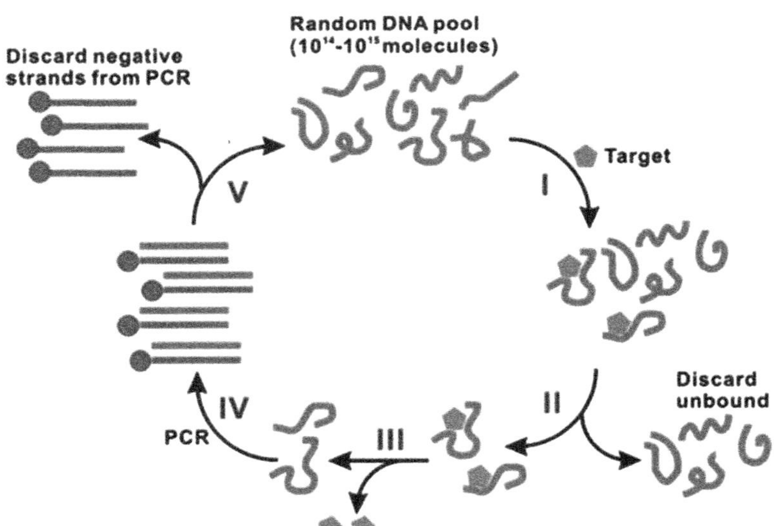

Fig. 11.1 In vitro selection of target-specific aptamers using SELEX technology (Reprinted with the permission from Ref. [12]. Copyright 2009 American Chemical Society)

are several properties of aptamer that make it an excellent nucleic acid probe of choice for interaction study. First, they bind to their target with good selectivity and affinity, and the binding affinity can be fine-tuned easily by site-specific mutations. Second, it is convenient to be synthesized by chemical methods in a large quantity and with perfect reproducibility. Third, aptamers are easy for conjugation with different fluorescent dyes, radiolabels, or other biomolecules. In addition, aptamer sequences are more stable than proteins under a wide range of conditions and could be repeatedly used without losing their binding capabilities. Moreover, aptamers are suitable for many unique signaling methods because of the predictable and tailorable structures of nucleic acids. It is possible to design aptamer probes based largely on their secondary structures with minimal knowledge of their tertiary structures [12]. A novel nucleic acid probe called molecular aptamer beacon (MAB) is developed by the excellent combination of the high binding affinity of aptamers and the sensitive signal transduction mechanism of MBs. In this chapter, different design approaches and applications of MABs will be introduced.

11.2 Designs of Molecular Aptamer Beacons

Similar to MBs, general MABs are designed by labeling a fluorophore and a quencher on either end of the stem. However, different from MBs, whose loop moiety is usually the complementary part of target, not only the loop moiety, even partial stem is able to be used as recognition elements. This special variation brings a diverse design mechanism into MABs.

11.2.1 Molecular Beacon Design

The most direct way to construct MABs is to replace the loop part of MBs to a target-specific sequence. Tan et al. have developed a MAB targeting to single-stranded DNA-binding protein to investigate the protein-DNA interaction [8]. In this design, the target molecule is protein which results in restoration of the fluorescence signal. In order to integrate MBs and aptamers and improve the binding affinity of MABs to target molecules, Ellington et al. have proposed a general method for direct selection of MAB [16]. A random pool of synthetic DNA sequences with fluorophore label close to $5'$ end is synthesized firstly. After annealing with a short piece of capture oligonucleotide labeled with quencher and biotin on $5'$ and $3'$ end, respectively, the capture pool is incubated with streptavidin-agarose and transferred to a column. There is no fluorescence emission due to the fluorescence resonance energy transfer between fluorophore and quencher. After addition of ligand, the fluorescence intensity of fractional ssDNA will restore, which demonstrates the binding event between ligands and ssDNA. All elutes are collected and amplified by PCR.

Successful selection of MABs has proved their potential in non-nucleic acid target classes, such as ions, small molecules, and even proteins. Takenaka et al.

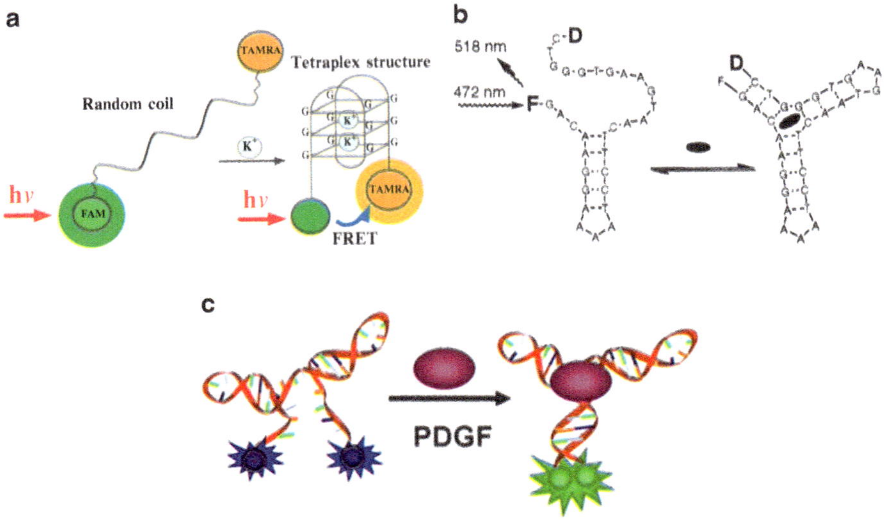

Fig. 11.2 Classical design of MABs. (**a**) Formation of G-quadruplex structure induced by potassium ions resulting in a FRET change (Reprinted with the permission from Ref. [17]. Copyright 2002 American Chemical Society). (**b**) Conformational change of anti-cocaine aptamer beacon in the presence of cocaine, resulting in a fluorescent quenching (Reprinted with the permission from Ref. [18]. Copyright 2001 American Chemical Society). (**c**) Use of the pyrene excimer to probe PDGF (Reprinted with the permission from Ref. [20]. Copyright 2005 National Academy of Sciences, USA)

constructed a potassium-sensing MAB for the specific detection K^+ in buffer solution [17]. Two different dyes, FAM and TAMRA, are labeled at the 5′- and 3′-termini of a potassium-specific aptamer beacon. In the presence of K^+, a fluorescence resonance energy transfer can be observed and used as a quantitative tool for the detection of potassium ions (Fig. 11.2a). The selectivity of MAB for K^+ is 43,000 times higher than Na^+. Another selected MAB targeting cocaine was reported by Stojanovic et al. [18]. A fluorophore and quencher were labeled on the either end of cocaine aptamer beacon (Fig. 11.2b). After addition of cocaine molecules, the MABs were able to fold to a three-way junction form and bring fluorophore and quencher closer, resulting in fluorescence intensity decrease. Even in the complex serum media, 10 μM cocaine was able to be detected. A similar fluorophore-quencher labeling strategy was reported by Ellington et al. to construct MABs for direct detection of thrombin [19]. Tan et al. have developed a novel dual-pyrene-labeled MAB, which are able to bind with high affinity to target protein platelet-derived growth factor (PDGF)-BB [20] (Fig. 11.2c). When the dual-pyrene-labeled aptamer beacon is free in solution without the target protein, both pyrene molecules are spatially separated, and only the monomer emission peaks (at 375 and 398 nm) are observed. The binding of the aptamer beacon to the target protein brings the pyrene molecules at 3′ and 5′ ends together, allowing the formation of an excimer. Thus, the emission peak at nearly 485 nm appears. The changes

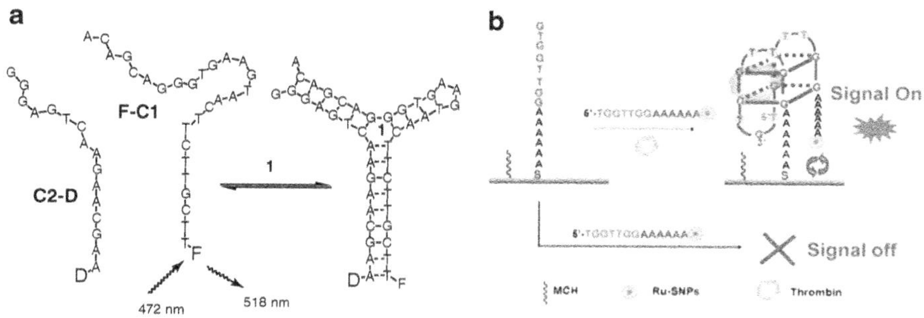

Fig. 11.3 Split MABs. (**a**) Self-assembly of split MABs in the presence of cocaine (Reprinted with the permission from Ref. [21]. Copyright 2000 American Chemical Society). (**b**) ECL sensor based on split thrombin aptamer subunits (Reproduced from Ref. [22] by permission of The Royal Society of Chemistry)

in emission color enabled visual detection of 4 pmol of PDGF in homogeneous solutions. Moreover, a strategy of time-resolved detection technique is adopted to eliminate most of the background fluorescence from the biological fluids, because the monomer and excimer emission of pyrene have much longer lifetimes (up to 100 ns) but most background fluorescence has a lifetime less than 5 ns.

11.2.2 Split Molecular Aptamer Beacons

Besides the intact MABs, other splitting MABs have been presented by dividing the MAB sequences into two subunits. In order to achieve such molecular engineering approach to split an intact MAB, the splitting site is crucial and hypothesized to not affect the binding between splitting MABs and targets. After addition of targets, two subunits of splitting MAB can be induced to assemble a target-MAB complex.

Stojanovic et al. applied two halves of anti-cocaine and anti-adenosine MABs to sense cocaine and adenosine, respectively [21]. 6-FAM fluorophore was labeled at the 5′-termini of one halve, and dabcyl quencher was modified at 3′-termini of another halve. Cocaine and adenosine molecules can recognize the ligand-binding pocket resulting in a formation of target MAB, and fluorescence quenching was observed (Fig. 11.3a). This splitting MAB can distinguish the concentration range from 10 μM to 1 mM ATP and 10 to 150 μM cocaine simultaneously. However, the binding affinity of the self-assembled MABs ($\sim$200 μM) decreased compared to the intact ones (20 μM). Willner et al. constructed self-assembled MABs by modifying CdSe/ZnS quantum dots (QDs) at 3′-termini of subunit 1 and Atto 590 dye 5′-termini of subunit 2. In the presence of cocaine, fluorescence resonance energy transfer will be stimulated to achieve the detection limit of 1 μM [23].

To develop a general excimer signaling approach for MABs and overcome the reduction of binding affinity by using splitting MABs, Yang et al. proposed an optimized molecular engineering strategy for light-switching excimer probes by using two target-induced complementary aptamer fragments (CAF) to replace a single-stranded aptamer sequence [24]. The MAB was split into two pieces by breaking the parent sequence at nonbinding pocket loop region, and a series of different cocaine aptamer fragment subunits were prepared by changing GC content at nonbinding pocket areas. To ensure the introduction of cocaine can bring two fragment sequences together, non-modified sequences were prepared, which was followed by a simple UV absorption measurement screening assay. This screening approach ensured the selection of the most optimized pair of fragment MABs, which had the best sensitivity and comparative binding affinity. The optimized complementary aptamer fragments were then labeled with pyrene molecule, one at 5′-termini and the other at 3′-termini. In the presence of target molecules, two complementary aptamer fragments were induced to assemble into an analyte-aptamer structure, resulting in the formation of pyrene excimer. With an anti-cocaine MAB, as low as 1 μM of cocaine can be detected using a fluorometer, and more importantly low picomole level of target can be directly visualized with naked eyes.

Chen et al. applied split MABs to construct electrochemiluminescence (ECL) biosensor for a highly sensitive detection of thrombin [22]. Antithrombin MAB was divided into two fragments: capture part modified with thiol and immobilized on the gold electrode and detection part functionalized with $Ru(bpy)_3{}^{2+}$-doped silica nanoparticles (Ru-SNPs). In the presence of thrombin, a simultaneous G-quadruplex structure was formed resulting in the observation of ECL (Fig. 11.3b). The dynamic range of thrombin concentration was 0.33–33 pM, and detection limit was 0.2 pM. Berger et al. investigated the binding affinity between small molecules and split MABs with atomic force spectroscopy (AFS) [25]. By monitoring the change of rupture force with and without addition of AMP, the value of Kd was determined comparative to the dissociation constant reported previously.

11.2.3 Structure-Switching Molecular Aptamer Beacons

Besides general and split MAB, there is another type of structure-switching signaling aptamers that report molecular recognition events by switching structures of DNA/target complex [26, 27]. The basic principle is that the structure-switching MAB maintains a "signal-off" state before target binding; however, after triggering by the addition of target molecules, the conformation of MAB will change and result in a corresponding signal change. This idea of structure-switching MAB was firstly reported by Li et al., who selected this novel MABs targeting to ATP and GTP in vitro [28]. The duplex structure of structure-switching MAB has three domains: (1) A modified aptamer (MAP) sequence contains stem-1, stem-2, and a toehold tag. Stem-2 and toehold tag are usually designed as a target-responsive aptamer sequence. (2) FDNA labeled with a fluorophore dye at 5′-termini is complementary to Stem-1. (3) QDNA modified with a quencher at 3′-termini hybridizes with Stem-2. This duplex structure keeps the fluorophore and quencher pair close to each other before adding target. In the presence of target, the binding between target and MAP will cause the structure switching and recover the fluorescence signal [26].

There are several variations from the basic design of structure-switching MAB: (1) Fluorophore is directly labeled at 5′-termini of a MAB, and QDNA modified with quencher is partially complementary to MAB. This design is relatively simple due to a duplex structure formed by double-stranded DNA sequences and suitable for short aptamer sequences. (2) Fluorophore is labeled in the middle of a MAB and quenched by a quencher in a close approach without addition of target. This design can be used for the binding site of MAB locating in the middle. (3) The whole sequence of a MAB is blocked by FDNA and QDNA, which is proper for long aptamer sequences. Such a variable design of structure-switching MABs provides a potential to be used for different types of aptamers [27].

Ho et al. developed fluorescent structure-switching MAB-based "signal-on" and "signal-off" mechanism [29]. Different from labeling fluorophore and quencher on either end of MAB, a fluorescent nucleotide analogue, such as 2-aminopurine and pyrrolo-dC, is introduced into competitor nucleic acid in the "signal-on" mode. The fluorescence will be strongly quenched by the base-pair stacking of MAB-competitor structure. After mixing with target, MAB and competitor will dehybridize, resulting in a strong restoration of fluorescence emission. Koide et al. have synthesized a photoinduced electron-transfer (PET) sensor with dichlorofluorescein and *N*-(*p*-methoxyphenyl)piperazine as fluorophore and quencher, respectively. RNA imaging was able to be achieved by using an RNA aptamer to specifically bind the quencher moiety of PET sensor, resulting in a 13-fold fluorescence enhancement [30].

11.3 Applications of Molecular Aptamer Beacons

11.3.1 Fluorescent Molecular Aptamer Beacon Sensors

Fluorophore labeling is the most principal approach in the signaling transduction of MABs because fluorophore possesses a large amount of advantages. Firstly, fluorescence has high sensitivity, which is important for the sample analysis with small volume or low analyte concentration. Secondly, fluorescence spectroscopy can achieve noninvasive and visual detection. Thirdly, by labeling a variety of fluorescent tags to different molecules, fluorescence spectroscopy allows multiplex detection and monitoring of various targets at the same time. Fourthly, the physical size of organic fluorophore is relatively small compared to biological tag, such as GFP, which may not be able to influence the binding affinity between MAB and target. Moreover, rich information about molecular interaction can be unveiled with a vast repertoire of fluorescence techniques including fluorescence anisotropy, fluorescence lifetime, fluorescence quenching, fluorescence wavelength switching, and fluorescence correlation spectroscopy [31].

Xu et al. have developed an aptamer-based exonuclease protection assay for the ultrasensitive detection of thrombin [32]. A nucleic acid probe contained one piece of thrombin MAB domain at the 3′-terminus bind to thrombin molecules, which could protect the whole nucleic acid sequence from degradation by exonuclease I. With the help of T4 ligase, the protected thrombin MAB will link two extra connector DNAs together as a quantitative PCR template for further signal amplification. The detection limit was 540 thrombin molecules, and linear range was from 540 to 540,000. Heyduk et al. designed two respective specific thrombin aptamers linking with signaling DNA sequences labeled with different fluorescence donor-accept pairs. After binding with thrombin simultaneously, these two signaling pieces hybridized to result in fluorescence resonance energy transfer (FRET). By optimizing the linker length and type of fluorescence donor-acceptor pair, 50 pM of thrombin was detected with high signal-to-background ratio [33]. As mentioned in Sect. 11.2.3, the fluorescence intensity of fluorescent nucleotide analogues varies as the conformation of MABs change with binding specific targets. Katilius et al. investigated the fluorescence enhancement of a series of fluorescent nucleotide analogues, such as 2-aminopurine (2AP), 4-amino-6-methylpteridone (6MAP), and 3-methylisoxanthopterin (3MI), incorporated into thrombin, IgE, and PDGF aptamers [34]. The result showed thrombin-modified fluorescent aptamer incorporated with 6MAP possessed 30-fold increase of fluorescence signal. Yu et al. demonstrated the use of MAB structure-switching-triggered ligation-rolling circle amplification (L-RCA) for the universal and high sensitive detection of protein. This approach can detect attomolar scale of PDGF [35]. Zhang et al. utilized the property of telomerase that produced telomere elongation reaction to open a hairpin DNA and monitor the telomerase activity [36]. Furthermore, with the help of polymerase for strand displacement, a cyclic signal amplification process can be achieved to measure telomerase activity equivalent to four cultured cancerous cells. Lu et al. constructed a novel catalytic molecular beacon (CAMB) by integrating the substrate of DNAzyme into the loop structure of a MB. Only in the presence of catalytic DNAzyme and corresponding cofactor, such as ions, the substrate of MB will be cleaved, resulting in the hybridization of a separated MB unstable. By monitoring the restoration of fluorescence, the detection limit of target cofactor ions can be determined as low as 600 pM [37].

In order to regulate the activity of aptamer sequences, Hamilton et al. have developed the targeted reversibly attenuated probe (TRAP)-contained attenuator, intervening antisense, and inactive aptamer or ribozyme. In the absence of regulatory nucleic acid, such as mRNA, the activity of aptamer was silenced. In the presence of regulatory DNA, aptamer will be activated to bind target due to the hybridization of regulatory DNA and intervening sequence [38]. Tan et al. analyzed variants of specific protein biomarker, platelet-derived growth factor (PDGF), with fluorescence-quencher-labeled MABs for the need of high throughput [39]. By optimizing the condition, such as temperature, pH, and cation concentrations of bioassays, as low as 10 ng PDGF was detected in the complex biological media [40]. An ultrasensitive detection of immunoglobulin E (IgE) was achieved by using IgE MAB with pyrene-labeled G-quadruplex and S1 nuclease, which were taken as signal transduction and amplification [41]. In the presence of IgE, MABs were

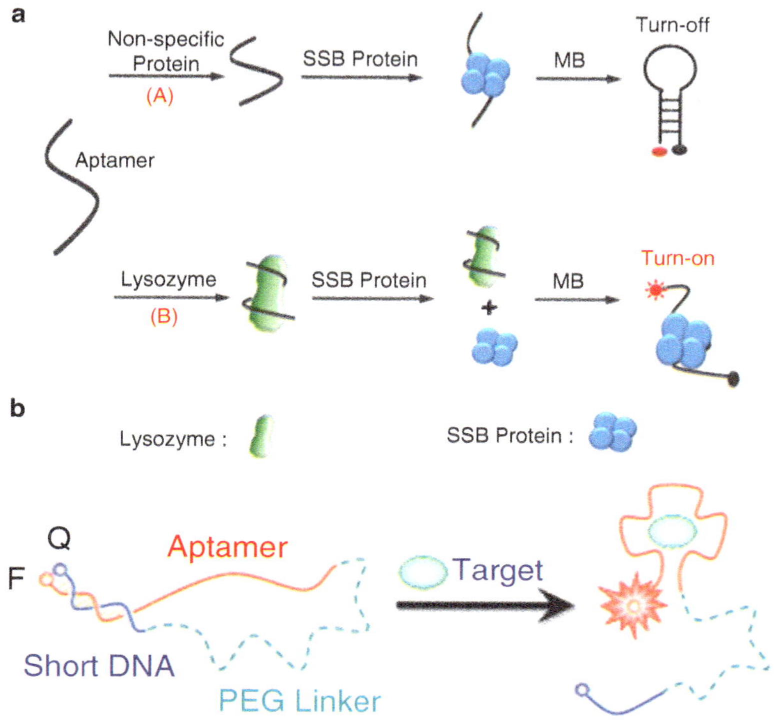

Fig. 11.4 (**a**) Fluorescence turn-on detection of lysozyme through the displaced single-stranded DNA-binding protein binding to a MAB (Reproduced from Ref. [42] by permission of The Royal Society of Chemistry). (**b**) Aptamer switch probe and design of ATP-ASP and human α-thrombin ASP (Tmb-ASP) (Reprinted with the permission from Ref. [44]. Copyright 2008 American Chemical Society)

able to bind the target, and signaling probes would assemble into G-quadruplex resulting in an emission fluorescence signal of pyrene excimer. After addition of S1 nuclease, the G-quadruplex structure would be cleaved to diminish the pyrene excimer emission. The detection limit of this method is 94.5 fM.

As mentioned in Sect. 11.2.1, single-stranded nucleic acids can specifically wrap around *E. coli*. Using this strategy, Yu et al. developed an approach for fluorescence turn-on detection of lysozyme with a label-free anti-lysozyme aptamer as binding element and a piece of MB targeting to SSB protein as signaling element (Fig. 11.4a). In the presence of lysozyme, the single-stranded aptamer will bind lysozyme and cannot wrap around SSB protein, which is able to open the loop-stem structure of additional MB, resulting in the increase of fluorescence signal. The detection limit of this method is 200 pM (2.86 ng/mL) [42]. Fluorophore-quencher-labeled thrombin MAB was also constructed to study real-time interaction between proteins [43]. The basic principle is to form thrombin-15 nt aptamer structure firstly, and no fluorescence emission observed due to FRET. After addition of a competition protein, such as HirF, binding to the same site as 15-nt antithrombin MAB, the fluorescence signal will be restored. Similarly, single fluorophore dye-labeled antithrombin aptamer and fluorescence anisotropy method are employed to investigate the protein-protein interaction. The calculating K_d of α-thrombin/HirF binding reaction is 180 ± 50 nM comparable to literature reported.

Kolpashchikov et al. have separated malachite green aptamer (MGA) into two fragments and attached two binding arms hybridizing for different nucleic acid analyte at each end of fragments. Only in the presence of target DNA analytes, these three strands can assemble into a tripartite complex which binds MG molecules again and results in over 20-fold enhancement of fluorescence signal [45]. Tan et al. modified MABs into a versatile intramolecular signal transduction aptamer probe, which contains an aptamer, a short DNA sequence complementary to part of the aptamer, and a PEG linker connecting these two [44]. FAM and Black Hole Quencher 2 are covalently attached at the two termini of the conjugated DNA sequences. In the absence of target molecule, the short complementary DNA will hybridize with partial aptamer thus keeping fluorophore and quencher in a close proximity. While after addition of target molecules, the conformation of aptamer switch probe will be changed due to the binding between aptamer and target and move the quencher away from the fluorophore. This novel aptamer switch probe is used to sense ATP and thrombin.

Different from fluorophore-quencher labeling method, Zhao et al. constructed thrombin MABs using guanine-quenching approaches [46]. Two strategies are employed: (1) fluorophore is labeled at one terminus of a hairpin structure of MAB and is quenched by the adjacent G bases at the termini. The conformation change is induced by the addition of thrombin which makes the adjacent G bases far away from fluorophore. (2) A double-stranded DNA is constructed, in which the binding site of aptamer to thrombin is blocked by a partial complementary DNA. In the presence of thrombin, the binding of MAB and thrombin will compete with the initial cDNA, resulting in the restoration of fluorescence signal. More functional proteins and domains have been investigated by MABs. Li et al. have constructed a MAB targeting myotonic dystrophy kinase-related Cdc42-binding kinase α (MRCK) [47]. Six additional bases complementary to $3'$-termini are added to fold the secondary structure into a hairpin structure, and fluorophore-quencher pair is labeled at the ends of hairpin structure. In the presence of MRCK, an MRCK-MAB complex is formed to signal the fluorescence. A 2.3-fold enhancement of fluorescence signal is observed by using MAB of MRCK, and the detection limit is 0.5 μM. Xie et al. developed a "signal-on" MB (807-35 nt) to detect recombinant human erythropoietin (rHuEPO-α) [48]. Fluorophore is labeled at the $3'$-termini of rHuEPO-α MAB (F-Apt), and a short DNA with quencher (QDNA) complementary to the stem part quenches the fluorescence. Another short-strand DNA (SDNA) stabilizes the secondary structure of hairpin MAB. In the presence of rHuEPO-α, QDNA is departed from F-Apt and restoration of fluorescence can be observed. This method is more sensitive (LOD: 0.4 nM) compared to "signal-off" method, whose LOD is 10 nM.

11.3.2 Electrochemical Molecular Aptamer Beacon Sensors

To overcome the disadvantages of fluorescent MAB sensors, for example, the real fluorescent signals of aptamer-target binding are easily overwhelmed by the background interference from surrounding environments, especially in the complex biological systems like cell media, serum, and blood, recently many design ideas of fluorescent aptamer sensors are adopted in the construction of electrochemical ones. Electrochemical approaches have presented many excited attributions over fluorescent MABs [49–52], such as high detection sensitivity, usually sub-nanomolar detection limits. More importantly, electrochemical biosensors of MABs are constructed on the surface of electrodes that rarely interfered by the complicated analytical environments, and target-responsive MABs is reagentless and reusable [53], so these sensors can be directly applied in complex condition conveniently.

Plaxco et al. firstly developed an electrochemical biosensor for sensitive DNA detection based on conformation-changed MBs [54]. Redoxable tag, ferrocence, was labeled at the 5′-termini of hairpin DNA, and thiol group was modified at 3′-termini to interact with gold electrode. In the absence of target DNA, the hairpin DNA would maintain the most stable loop-stem structure to achieve high electron-transfer efficiency; while in the presence of target DNA, hybridization of target DNA and MB would induce the ferrocence tag far away from the surface of gold electrode, resulting in a decrease of electron-transfer efficiency. By monitoring the change of electron-transfer efficiency, down to 10 pM, target DNA can be measured by cyclic voltammetry method. This design of electrochemical sensor opens a gate for analysis of real sample in complex condition with high sensitivity and selectivity. Grinstaff et al. constructed a "signal-on" electrochemical DNA sensor by using two short-strand DNAs: one is capture part and the other is signaling part that is linked with a flexible PEG spacer [55]. The hybridization of target DNAs to the triblock DNA probe will induce the conformation change that results in the close proximity of electrochemical reporter to the surface of gold electrode. This change of conformation brings electrochemical "signal-on" amplification that can be used as a quantitative analysis of target DNAs. The method shows that the detection limit for prototype DNA is ~200 pM. Compared to fluorescent MAB sensors, these electrochemical sensors are more reagentless and reusable. O'Sullivan et al. developed surface-attached MABs combined with electrochemical approach to analyze protein durably [56]. Two unique points are mentioned: firstly, 2-mercaptoethanol is used to block electrode surface and discriminate the unbound thrombin from gold electrode; secondly, this electrochemical MAB platform can be regenerated by immersing MABs into hydrochloric acid solution for 25 times without losing the electrochemical signal and binding affinity to thrombin. Plaxco et al. modified this electrochemical "signal-on" MAB sensor into a target-induced strand displacement mechanism to directly exploiting target-responsive conformation change [57]. Hybridization of target DNA displaces signaling part resulting in the electrochemical tag, methylene blue, and close proximity to surface of Au electrode, and an increase electron transfer is observed. The limit of detection is down to 400 fM.

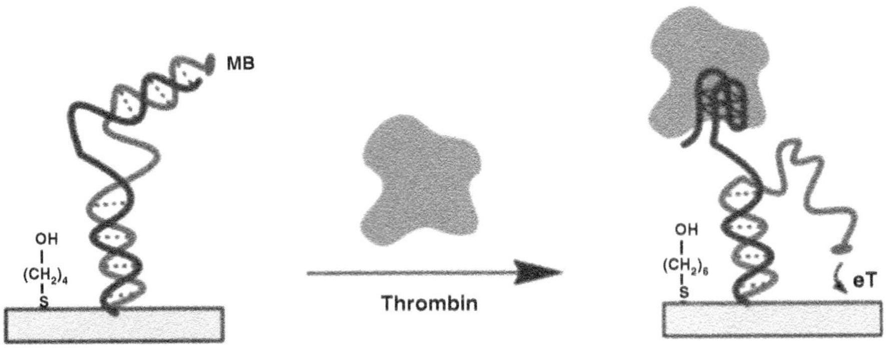

Fig. 11.5 "Signal-on" electrochemical biosensor for sensitive thrombin detection based on target-induced strand displacement (Reprinted with the permission from Ref. [60]. Copyright 2005 American Chemical Society)

Plaxco et al. firstly proposed a "signal-off" electronic MAB sensor for the detection of thrombin [58]. This sensor was also proven suitable for blood serum. Different from covalent labeling of electronic tag at the end of MABs, Kim et al. developed an electrochemical detection of thrombin by employing MABs with intercalation tag [59]. In the presence of thrombin, due to the conformation change of MAB binding with thrombin, intercalated methylene blue tags were released, resulting in a decrease of current intensity. The detection limit of this method was 11 nM of thrombin. To solve the limitation of signal-off electronic MAB sensors that initial signal cannot be suppressed well, a double-stranded DNA was used to construct signal-on MAB sensor, in which DNA 1 containing thrombin aptamer was covalently linked on the surface of gold electrode and DNA 2 modified with methylene blue was hybridized with DNA 1 to make the complementary structure rigid and reduce background signals [60]. Upon binding with thrombin, DNA 2 was displaced to liberate MB tag to be close to the surface of gold electrode, resulting in detectable electronic signals (Fig. 11.5). This method improved the detection limit of thrombin to as low as 3 nM. Carminic acid (CA)-labeled electrochemically active-inactive switch aptamer molecular beacon was developed by Cheng et al. [61]. In the absence of thrombin, CA dimer was formed due to the hairpin structure of MAB, resulting in inactive electrochemical signals. While, in the presence of thrombin, binding of 5′-termini of MAB and thrombin dissociated CA dimer to form active CA monomer. By enhancing the electrochemical signals with magnetic nanobeads, the detection limit of this method can achieve as low as 42.4 pM. The similar design of electrochemical MAB sensor was constructed for the detection of PDGF-B directly in the undiluted blood serum [62]. This method achieved a higher sensitivity and selectivity against optical sensors, reaching a 50pM detection limit in serum-diluted condition.

Besides electrochemical MAB sensors for the detection of proteins, Plaxco et al. also used anti-cocaine MABs to construct electronic sensor [63]. Only the electronic active tag methylene blue (MB) labeled at 3′-termini of the MAB can produce a high faradaic currents change upon binding with cocaine targets. Fan et al. developed a target-responsive electrochemical aptamer switch (TREAS) approach for the detection of ATP [64]. This method constructed a duplex DNA structure on gold electrodes, in which one is anti-ATP MAB modified with ferrocene tag and the other one is hybridizing DNA. In the absence of ATP, the ferrocene tag was far away from gold electrode that had no faradaic currents produced. However, binding of ATP can induce the conformation change of ATP MAB, and a 10-fold signal gain was observed in the presence of ATP. Although RNA MABs are easily affected by the risk of RNase contamination, the first electrochemical sensor of RNA MAB was presented by Ferapontova et al. for the detection of theophylline in serum [65]. After binding with theophylline, RNA MAB folded into two internal loop conformations resulting in an increased electron-transfer efficiency. The dynamic range and detection limit were 0.2–10 μM and 0.2 μM, respectively. DNAzyme was incorporated on the surface of gold electrode to construct lead recognition sensors [66]. An electronic tag, MB, labeled catalytic DNA strand was modified on the Au electrode by Au-S bonding; the other substrate DNA was hybridized to form a DNAzyme complex. With the help of lead ions, the substrate was cleaved resulting in MB close proximity to gold electrode. After optimization, this method can detect as low as 0.3 μM, equivalently 62 ppb Pb^{2+}. Besides, i-motif quadruplex MABs are able to sense down to 0.2 ppm single-walled carbon nanotubes (SWNTs) due to SWNTs are beneficial for the formation of i-motif structure in buffer solution [67]. Soh et al. reported a MAB-based electrochemical sensor built on microfluidic chips (MECAS) for continuous, real-time, and high-throughput monitoring of cocaine in blood serum directly [68]. This microfluidic sensor has a high reproducibility when reusing in the detection of micromolar level of cocaine.

11.3.3 Optical Molecular Aptamer Beacon Sensors

Optical (colorimetric) MABs possess many advantages compared to fluorescent and electrochemical MAB sensors, such as facile and fast visual detection of target molecules without any complicated pretreatment process under room temperature. Of course, these simple optical methods suffer from the limitation of relatively low sensitivity in comparison to spectroscopic and electrochemical approaches. There are two major strategies in the design of optical MAB sensors based on gold nanoparticles (AuNPs). The first one is AuNP-DNA conjugates that Mirkin's group introduced and developed [72]. The other one is to use the interaction between unmodified aptamer beacons and gold nanoparticles to construct biosensors.

Fan's group developed a series of multicolor nanobeacons and nanoprobes by conjugating different types of MBs and MABs labeled with various dyes on the surface of Au nanoparticles for multiple sensing of tumor DNA markers, small molecules, and proteins [73, 74]. Stojanovic et al. selected a cyanine dye,

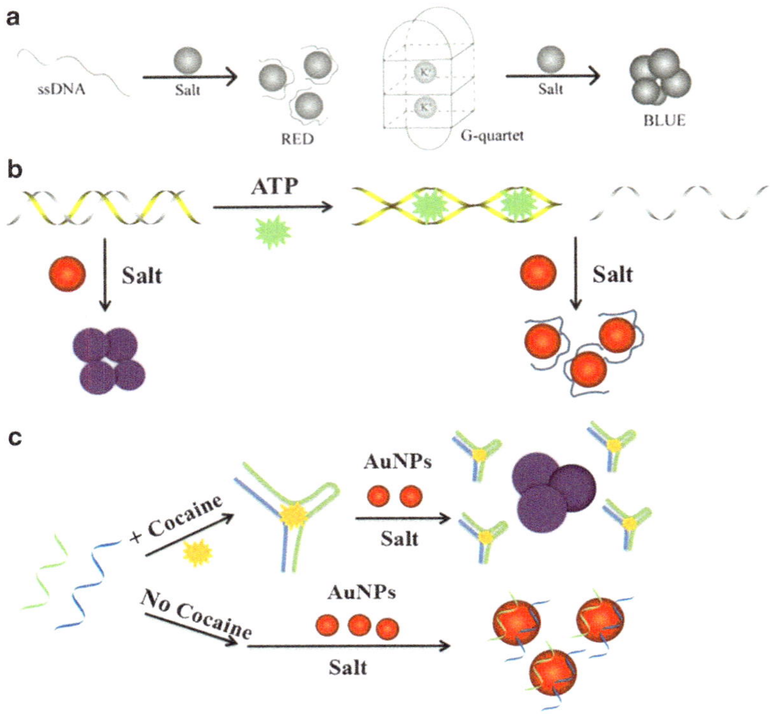

Fig. 11.6 Optical biosensors based on MABs developed by Fan's group. (**a**) Reproduced from Ref. [69] by permission of The Royal Society of Chemistry. (**b**) Reproduced from Ref. [70] by permission of John Wiley & Sons Ltd. (**c**) Reproduced from Ref. [71] by permission of John Wiley & Sons Ltd.

diethylthiotricarbocyanine iodide, to construct an optical MAB for cocaine sensing [75]. Before addition of cocaine, the cyanine dye can intercalate into the loop pocket of anti-cocaine aptamer and show a visible absorbance; while in the presence of cocaine, cyanine dye was displaced resulting in an obvious attenuation of absorbance. The sensitivity of this colorimetric MAB is 0.5 μM. Fan et al. also constructed a series of colorimetric gold nanoparticles sensors by using MABs. For example, as shown in Fig. 11.6a, a single-stranded potassium MAB was employed to prevent the aggregation of AuNPs in the absence of potassium ions. While, the blue color of aggregated AuNPs was observed due to the formation of potassium-MAB complex [69]. Similar for ATP MAB, i-motif, and mercury-specific oligonucleotide (MSO), the same strategy was employed to construct colorimetric sensors (Fig. 11.6b) [70, 76]. Not all MABs undergo a large target-responsive conformation change when sensed by AuNPs [71], so the split MABs were used to sense cocaine with unmodified AuNPs (Fig. 11.6c). With this approach, as low as 2 μM cocaine can be distinguished by the naked eye. Wang et al. exploited water-soluble conjugated polymer, PMNT, and mercury-specific oligonucleotide probe for direct visual

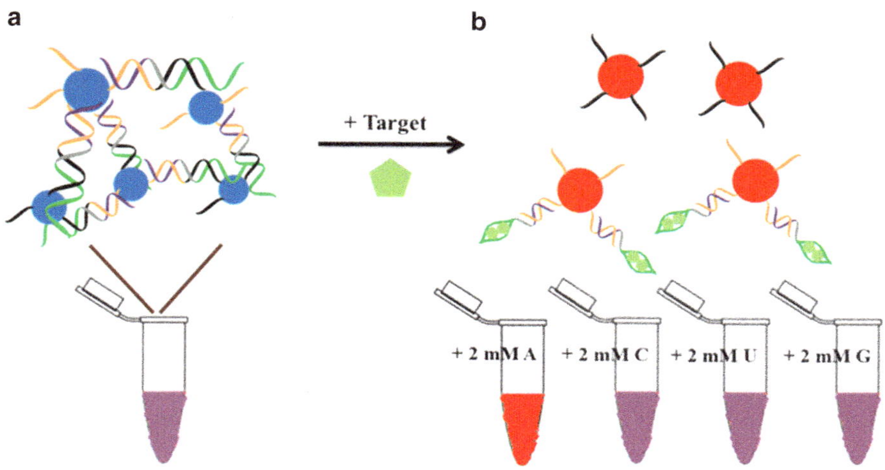

Fig. 11.7 Scheme of colorimetric detection of ATP based on target-induced disassembly of gold nanoparticles aggregates (**a**) and visual result after addition of adenosine or other nucleosides, cytidine, uridine, guanosine (**b**) (Reproduced from Ref. [78] by permission of John Wiley & Sons Ltd.)

detection of mercury ions. In the presence of Hg^{2+}, T-Hg-T structure was formed to interact with PMNT and keep the yellow color of conjugated polymer. In the absence of Hg^{2+}, a MSO-PMNT complex was formed and displayed a red color. By identifying this obvious yellow-to-red color change by the naked eye, 12.5 μM Hg^{2+} can be detected. The detection limit could be lowered to 2.5 μM with UV-vis spectrometry [77]. Lu et al. developed a colorimetric detection of ATP and cocaine based on target-induced disassembly of gold nanoparticles aggregates (Fig. 11.7) [78]. Two types of short DNA sequences modifying on AuNPs were hybridized with linker MAB, which brought AuNPs aggregation. In the presence of targets, this aggregated AuNPs would dissociate and show an obvious color change due to the binding of adenosine or cocaine and the linker MABs. 0.3 ~ 2 mM ATP and 50 ~ 500 μM cocaine can be visually detected by the naked eye with this approach. The similar strategy was used to construct colorimetric AuNPs sensors based on MABs for the detection of mercury ions [79–82] and proteins, such as PDGF [83] and thrombin [84].

Deng et al. reported anti-adenosine MAB assembled with fragmented hemin-binding DNAzyme for the visual detection of ATP. In the absence of adenosine, the split DNAzyme assembled to form G-quadruplex structure with the help of hemin and K^+, further catalyzing $ABTS^{2+}$ to a blue-green-colored radical product, $ABTS^{\bullet+}$. In the presence of adenosine, the G-quadruplex was not formed due to the binding of ATP and its MABs. A limit of detection of 6 μM is achieved by this method [85]. Willner et al. also constructed the aptamer-DNAzyme hairpins for the biosensing of small molecules, like AMP and proteins, such as lysozyme [86].

In the absence of targets, the horseradish peroxidase (HRP)-mimicking DNAzyme sequence was protected in the stem of hairpin structure and kept an inactive state. While, in the presence of targets, G-quadruplex structure of DNAzyme was formed due to the binding of targets and MABs. The detection limits are 50 μM and 0.5 pM for AMP and lysozyme, respectively.

11.4 Conclusion and Prospective

Molecular aptamer beacons combine high selectivity of aptamers and conformation switch of MBs to perform a variety of biological applications, like enzyme monitoring [87–89], mRNA detection in living cells [90–92], and biosensors [5, 6, 93, 94]. This chapter centered on the angles of molecular recognition and sensing integrated with fluorescence, electrochemistry, and nanoparticles. Besides, MABs combining with single-walled carbon nanotubes are developed to produce singlet oxygen for photodynamic therapy [95, 96]. In the future, MABs will find their great potential in the practical applications, such as high-throughput drug screening, disease diagnosis, nanotechnology, and materials science.

References

1. Tyagi S, Kramer FR (1996) Molecular beacons: probes that fluoresce upon hybridization. Nat Biotechnol 14(3):303–308
2. Tyagi S, Bratu DP, Kramer FR (1998) Multicolor molecular beacons for allele discrimination. Nat Biotechnol 16(1):49–53
3. Tyagi S, Marras SAE, Kramer FR (2000) Wavelength-shifting molecular beacons. Nat Biotechnol 18(11):1191–1196
4. Zhang P, Beck T, Tan W (2001) Design of a molecular beacon DNA probe with two fluorophores. Angew Chem Int Ed 40(2):402–405
5. Fang X, Liu X, Schuster S, Tan W (1999) Designing a novel molecular beacon for surface-immobilized DNA hybridization studies. J Am Chem Soc 121(12):2921–2922
6. Li J, Tan W, Wang K, Xiao D, Yang X, He X, Tang Z (2001) Ultrasensitive optical DNA biosensor based on surface immobilization of molecular beacon by a bridge structure. Anal Sci 17(10):1149–1153
7. Fang X, Li JJ, Tan W (2000) Using molecular beacons to probe molecular interactions between lactate dehydrogenase and single-stranded DNA. Anal Chem 72(14):3280–3285
8. Li JJ, Fang X, Schuster SM, Tan W (2000) Molecular beacons: a novel approach to detect protein – DNA interactions. Angew Chem Int Ed 39(6):1049–1052
9. Tan W, Fang X, Li J, Liu X (2000) Molecular beacons: a novel DNA probe for nucleic acid and protein studies. Chem Eur J 6(7):1107–1111
10. Li JJ, Fang X, Tan W (2002) Molecular aptamer beacons for real-time protein recognition. Biochem Biophys Res Commun 292(1):31–40
11. Cao Z, Suljak SW, Tan W (2005) Molecular beacon aptamers for protein monitoring in real-time and in homogeneous solutions. Curr Proteomics 2:31–40
12. Liu J, Cao Z, Lu Y (2009) Functional nucleic acid sensors. Chem Rev 109(5):1948–1998

13. Ellington AD, Szostak JW (1990) In vitro selection of RNA molecules that bind specific ligands. Nature 346(6287):818–822
14. Robertson DL, Joyce GF (1990) Selection in vitro of an RNA enzyme that specifically cleaves single-stranded DNA. Nature 344(6265):467–468
15. Tuerk C, Gold L (1990) Systematic evolution of ligands by exponential enrichment: RNA ligands to bacteriophage T4 DNA polymerase. Science 249(4968):505–510
16. Rajendran M, Ellington AD (2003) In vitro selection of molecular beacons. Nucleic Acids Res 31(19):5700–5713
17. Ueyama H, Takagi M, Takenaka S (2002) A novel potassium sensing in aqueous media with a synthetic oligonucleotide derivative. Fluorescence resonance energy transfer associated with guanine quartet – potassium ion complex formation. J Am Chem Soc 124(48):14286–14287
18. Stojanovic MN, de Prada P, Landry DW (2001) Aptamer-based folding fluorescent sensor for cocaine. J Am Chem Soc 123(21):4928–4931
19. Hamaguchi N, Ellington A, Stanton M (2001) Aptamer beacons for the direct detection of proteins. Anal Biochem 294(2):126–131
20. Yang CJ, Jockusch S, Vicens M, Turro NJ, Tan W (2005) Light-switching excimer probes for rapid protein monitoring in complex biological fluids. Proc Natl Acad Sci USA 102(48):17278–17283
21. Stojanovic MN, de Prada P, Landry DW (2000) Fluorescent sensors based on aptamer self-assembly. J Am Chem Soc 122(46):11547–11548
22. Lin Z, Chen L, Zhu X, Qiu B, Chen G (2010) Signal-on electrochemiluminescence biosensor for thrombin based on target-induced conjunction of split aptamer fragments. Chem Commun (Camb) 46(30):5563–5565
23. Freeman R, Li Y, Tel-Vered R, Sharon E, Elbaz J, Willner I (2009) Self-assembly of supramolecular aptamer structures for optical or electrochemical sensing. Analyst 134(4):653–656
24. Wu C, Yan L, Wang C, Lin H, Wang C, Chen X, Yang CJ (2010) A general excimer signaling approach for aptamer sensors. Biosens Bioelectron 25(10):2232–2237
25. Nguyen T-H, Steinbock LJ, H-Jr B, Helm M, Rd B (2011) Measuring single small molecule binding via rupture forces of a split aptamer. J Am Chem Soc 133(7):2025–2027
26. Nutiu R, Li Y (2003) Structure-switching signaling aptamers. J Am Chem Soc 125(16):4771–4778
27. Nutiu R, Li Y (2004) Structure-switching signaling aptamers: transducing molecular recognition into fluorescence signaling. Chem Eur J 10(8):1868–1876
28. Nutiu R, Li Y (2005) In vitro selection of structure-switching signaling aptamers. Angew Chem Int Ed 44(7):1061–1065
29. Li N, Ho C-M (2008) Aptamer-based optical probes with separated molecular recognition and signal transduction modules. J Am Chem Soc 130(8):2380–2381
30. Sparano BA, Koide K (2005) A strategy for the development of small-molecule-based sensors that strongly fluoresce when bound to a specific RNA. J Am Chem Soc 127(43):14954–14955
31. Wu C, Wang C, Yan L, Yang CJ (2009) Pyrene excimer nucleic acid probes for biomolecule signaling. J Biomed Nanotechnol 5(5):495–504
32. Wang X-L, Li F, Su Y-H, Sun X, Li X-B, Schluesener HJ, Tang F, Xu S-Q (2004) Ultrasensitive detection of protein using an aptamer-based exonuclease protection assay. Anal Chem 76(19):5605–5610
33. Heyduk E, Heyduk T (2005) Nucleic acid-based fluorescence sensors for detecting proteins. Anal Chem 77(4):1147–1156
34. Katilius E, Katiliene Z, Woodbury NW (2006) Signaling aptamers created using fluorescent nucleotide analogues. Anal Chem 78(18):6484–6489
35. Wu Z-S, Zhang S, Zhou H, Shen G-L, Yu R (2010) Universal aptameric system for highly sensitive detection of protein based on structure-switching-triggered rolling circle amplification. Anal Chem 82(6):2221–2227

36. Ding C, Li X, Ge Y, Zhang S (2010) Fluorescence detection of telomerase activity in cancer cells based on isothermal circular strand-displacement polymerization reaction. Anal Chem 82(7):2850–2855
37. Zhang X-B, Wang Z, Xing H, Xiang Y, Lu Y (2010) Catalytic and molecular beacons for amplified detection of metal ions and organic molecules with high sensitivity. Anal Chem 82(12):5005–5011
38. Cong X, Nilsen-Hamilton M (2005) Allosteric aptamers: targeted reversibly attenuated probes†. Biochemistry 44(22):7945–7954
39. Fang X, Sen A, Vicens M, Tan W (2003) Synthetic DNA aptamers to detect protein molecular variants in a high-throughput fluorescence quenching assay. Chembiochem 4(9):829–834
40. Vicens MC, Sen A, Vanderlaan A, Drake TJ, Tan W (2005) Investigation of molecular beacon aptamer-based bioassay for platelet-derived growth factor detection. Chembiochem 6(5):900–907
41. Wu Z-S, Hu P, Zhou H, Shen G, Yu R (2010) Fluorescent oligonucleotide probe based on G-quadruplex scaffold for signal-on ultrasensitive protein assay. Biomaterials 31(7):1918–1924
42. Tang D, Liao D, Zhu Q, Wang F, Jiao H, Zhang Y, Yu C (2011) Fluorescence turn-on detection of a protein through the displaced single-stranded DNA binding protein binding to a molecular beacon. Chem Commun (Camb) 47(19):5485–5487
43. Cao Z, Tan W (2005) Molecular aptamers for real-time protein–protein interaction study. Chem Eur J 11(15):4502–4508
44. Tang Z, Mallikaratchy P, Yang R, Kim Y, Zhu Z, Wang H, Tan W (2008) Aptamer switch probe based on intramolecular displacement. J Am Chem Soc 130(34):11268–11269
45. Kolpashchikov DM (2005) Binary malachite green aptamer for fluorescent detection of nucleic acids. J Am Chem Soc 127(36):12442–12443
46. Wang W, Chen C, Qian MX, Zhao XS (2008) Aptamer biosensor for protein detection based on guanine-quenching. Sensors Actuators B-Chem 129(1):211–217
47. Tok J, Lai J, Leung T, Li SFY (2010) Molecular aptamer beacon for myotonic dystrophy kinase-related Cdc42-binding kinase [alpha]. Talanta 81(1–2):732–736
48. Zhang Z, Guo L, Tang J, Guo X, Xie J (2009) An aptameric molecular beacon-based "Signal-on" approach for rapid determination of rHuEPO-[alpha]. Talanta 80(2):985–990
49. Odenthal KJ, Gooding JJ (2007) An introduction to electrochemical DNA biosensors. Analyst 132(7):603–610
50. Lee T (2008) Over-the-counter biosensors: past, present, and future. Sensors-basel 8(9):5535–5559
51. Mendes PM (2008) Stimuli-responsive surfaces for bio-applications. Chem Soc Rev 37(11):2512–2529
52. Privett BJ, Shin JH, Schoenfisch MH (2008) Electrochemical sensors. Anal Chem 80(12):4499–4517
53. Lubin AA, Lai RY, Baker BR, Heeger AJ, Plaxco KW (2006) Sequence-specific, electronic detection of oligonucleotides in blood, soil, and foodstuffs with the reagentless, reusable E-DNA sensor. Anal Chem 78(16):5671–5677
54. Fan C, Plaxco KW, Heeger AJ (2003) Electrochemical interrogation of conformational changes as a reagentless method for the sequence-specific detection of DNA. Proc Natl Acad Sci USA 100(16):9134–9137
55. Immoos CE, Lee SJ, Grinstaff MW (2004) DNA-PEG-DNA triblock macromolecules for reagentless DNA detection. J Am Chem Soc 126(35):10814–10815
56. Radi A-E, Acero Sánchez JL, Baldrich E, O'Sullivan CK (2005) Reagentless, reusable, ultrasensitive electrochemical molecular beacon aptasensor. J Am Chem Soc 128(1):117–124
57. Xiao Y, Lubin AA, Baker BR, Plaxco KW, Heeger AJ (2006) Single-step electronic detection of femtomolar DNA by target-induced strand displacement in an electrode-bound duplex. Proc Natl Acad Sci USA 103(45):16677–16680
58. Xiao Y, Lubin AA, Heeger AJ, Plaxco KW (2005) Label-free electronic detection of thrombin in blood serum by using an aptamer-based sensor. Angew Chem Int Ed 44(34):5456–5459

59. Bang GS, Cho S, Kim B-G (2005) A novel electrochemical detection method for aptamer biosensors. Biosens Bioelectron 21(6):863–870
60. Xiao Y, Piorek BD, Plaxco KW, Heeger AJ (2005) A reagentless signal-on architecture for electronic, aptamer-based sensors via target-induced strand displacement. J Am Chem Soc 127(51):17990–17991
61. Cheng G, Shen B, Zhang F, Wu J, Xu Y, He P, Fang Y (2010) A new electrochemically active-inactive switching aptamer molecular beacon to detect thrombin directly in solution. Biosens Bioelectron 25(10):2265–2269
62. Lai RY, Plaxco KW, Heeger AJ (2006) Aptamer-based electrochemical detection of picomolar platelet-derived growth factor directly in blood serum. Anal Chem 79(1):229–233
63. Baker BR, Lai RY, Wood MS, Doctor EH, Heeger AJ, Plaxco KW (2006) An electronic, aptamer-based small-molecule sensor for the rapid, label-free detection of cocaine in adulterated samples and biological fluids. J Am Chem Soc 128(10):3138–3139
64. Zuo X, Song S, Zhang J, Pan D, Wang L, Fan C (2007) A target-responsive electrochemical aptamer switch (TREAS) for reagentless detection of nanomolar ATP. J Am Chem Soc 129(5):1042–1043
65. Ferapontova EE, Olsen EM, Gothelf KV (2008) An RNA aptamer-based electrochemical biosensor for detection of theophylline in serum. J Am Chem Soc 130(13):4256–4258
66. Xiao Y, Rowe AA, Plaxco KW (2006) Electrochemical detection of parts-per-billion lead via an electrode-bound DNAzyme assembly. J Am Chem Soc 129(2):262–263
67. Peng Y, Wang X, Xiao Y, Feng L, Zhao C, Ren J, Qu X (2009) i-Motif quadruplex DNA-based biosensor for distinguishing single- and multiwalled carbon nanotubes. J Am Chem Soc 131(38):13813–13818
68. Swensen JS, Xiao Y, Ferguson BS, Lubin AA, Lai RY, Heeger AJ, Plaxco KW, Soh HT (2009) Continuous, real-time monitoring of cocaine in undiluted blood serum via a microfluidic, electrochemical aptamer-based sensor. J Am Chem Soc 131(12):4262–4266
69. Wang L, Liu X, Hu X, Song S, Fan C (2006) Unmodified gold nanoparticles as a colorimetric probe for potassium DNA aptamers. Chem Commun (Camb) 36:3780–3782
70. Wang J, Wang L, Liu X, Liang Z, Song S, Li W, Li G, Fan C (2007) A gold nanoparticle-based aptamer target binding readout for ATP assay. Adv Mater 19(22):3943–3946
71. Zhang J, Wang L, Pan D, Song S, Boey FYC, Zhang H, Fan C (2008) Visual cocaine detection with gold nanoparticles and rationally engineered aptamer structures. Small 4(8):1196–1200
72. Elghanian R, Storhoff JJ, Mucic RC, Letsinger RL, Mirkin CA (1997) Selective colorimetric detection of polynucleotides based on the distance-dependent optical properties of gold nanoparticles. Science 277(5329):1078–1081
73. Song S, Liang Z, Zhang J, Wang L, Li G, Fan C (2009) Gold-nanoparticle-based multicolor nanobeacons for sequence-specific DNA analysis. Angew Chem Int Ed 121(46):8826–8830
74. Zhang J, Wang L, Zhang H, Boey F, Song S, Fan C (2010) Aptamer-based multicolor fluorescent gold nanoprobes for multiplex detection in homogeneous solution. Small 6(2):201–204
75. Stojanovic MN, Landry DW (2002) Aptamer-based colorimetric probe for cocaine. J Am Chem Soc 124(33):9678–9679
76. Wang L, Zhang J, Wang X, Huang Q, Pan D, Song S, Fan C (2008) Gold nanoparticle based optical probes for target-responsive DNA structures. Gold Bull 41(1):37–41
77. Liu X, Tang Y, Wang L, Zhang J, Song S, Fan C, Wang S (2007) Optical detection of mercury(II) in aqueous solutions by using conjugated polymers and label-free oligonucleotides. Adv Mater 19(13):1662–1662
78. Liu J, Lu Y (2006) Fast colorimetric sensing of adenosine and cocaine based on a general sensor design involving aptamers and nanoparticles. Angew Chem Int Ed 45(1):90–94
79. Ono A, Togashi H (2004) Highly selective oligonucleotide-based sensor for mercury(II) in aqueous solutions. Angew Chem Int Ed 43(33):4300–4302
80. Tanaka Y, Oda S, Yamaguchi H, Kondo Y, Kojima C, Ono A (2006) 15N–15N J-coupling across HgII: direct observation of HgII-mediated T–T base pairs in a DNA duplex. J Am Chem Soc 129(2):244–245

81. Lee J-S, Han MS, Mirkin CA (2007) Colorimetric detection of mercuric ion (Hg^{2+}) in aqueous media using DNA-functionalized gold nanoparticles. Angew Chem Int Ed 46(22):4093–4096
82. Xue X, Wang F, Liu X (2008) One-step, room temperature, colorimetric detection of mercury (Hg^{2+}) using DNA/nanoparticle conjugates. J Am Chem Soc 130(11):3244–3245
83. Huang C-C, Huang Y-F, Cao Z, Tan W, Chang H-T (2005) Aptamer-modified gold nanoparticles for colorimetric determination of platelet-derived growth factors and their receptors. Anal Chem 77(17):5735–5741
84. Pavlov V, Xiao Y, Shlyahovsky B, Willner I (2004) Aptamer-functionalized Au nanoparticles for the amplified optical detection of thrombin. J Am Chem Soc 126(38):11768–11769
85. Lu N, Shao C, Deng Z (2008) Rational design of an optical adenosine sensor by conjugating a DNA aptamer with split DNAzyme halves. Chem Commun (Camb) 46:6161–6163
86. Teller C, Shimron S, Willner I (2009) Aptamer – DNAzyme hairpins for amplified biosensing. Anal Chem 81(21):9114–9119
87. Li JJ, Geyer R, Tan W (2000) Using molecular beacons as a sensitive fluorescence assay for enzymatic cleavage of single-stranded DNA. Nucleic Acids Res 28(11):e52
88. Tang Z, Wang K, Tan W, Li J, Liu L, Guo Q, Meng X, Ma C, Huang S (2003) Real-time monitoring of nucleic acid ligation in homogenous solutions using molecular beacons. Nucleic Acids Res 31(23):e148
89. Tang Z, Wang K, Tan W, Ma C, Li J, Liu L, Guo Q, Meng X (2003) Real-time investigation of nucleic acids phosphorylation process using molecular beacons. Nucleic Acids Res 33(11):e97
90. Bratu DP, Cha B-J, Mhlanga MM, Kramer FR, Tyagi S (2003) Visualizing the distribution and transport of mRNAs in living cells. Proc Natl Acad Sci USA 100(23):13308–13313
91. Medley CD, Drake TJ, Tomasini JM, Rogers RJ, Tan W (2005) Simultaneous monitoring of the expression of multiple genes inside of single breast carcinoma cells. Anal Chem 77(15):4713–4718
92. Mhlanga MM, Vargas DY, Fung CW, Kramer FR, Tyagi S (2005) tRNA-linked molecular beacons for imaging mRNAs in the cytoplasm of living cells. Nucleic Acids Res 33(6):1902–1912
93. Liu X, Tan W (1999) A fiber-optic evanescent wave DNA biosensor based on novel molecular beacons. Anal Chem 71(22):5054–5059
94. Steemers FJ, Ferguson JA, Walt DR (2000) Screening unlabeled DNA targets with randomly ordered fiber-optic gene arrays. Nat Biotechnol 18(1):91–94
95. Tang Z, Zhu Z, Mallikaratchy P, Yang R, Sefah K, Tan W (2010) Aptamer–target binding triggered molecular mediation of singlet oxygen generation. Asian J Chem 5(4):783–786
96. Zhu Z, Tang Z, Phillips JA, Yang R, Wang H, Tan W (2008) Regulation of singlet oxygen generation using single-walled carbon nanotubes. J Am Chem Soc 130(33):10856–10857

The manufacturer's authorised representative in the EU is Springer Nature Customer Service Centre GmbH, Europaplatz 3, 69115 Heidelberg, Germany. If you have any concerns regarding our products, please contact ProductSafety@springernature.com

Printed and bound by CPI Group (UK) Ltd, Croydon, CR0 4YY
29/07/2026
02180623-0001